Environmental Politics and Theory

The premise of this series is that the current environmental crisis cannot be solved by technological innovation alone. The environmental challenges we face today are, at their root, political crises involving political values, institutions and struggles for power. Therefore, environmental politics and theory are of the utmost social significance.

Growing public consciousness of the environmental crisis and its human and more-than-human impacts, exemplified by the worldwide urgency and political activity associated with the problem and consequences of climate and earth system change make it imperative to design and achieve a sustainable and socially just society.

The series publishes inter- and multi-disciplinary scholarship that extends the theoretical dimensions of green political theory, international relations, philosophy, and earth system governance. It addresses the need for social change away from the hegemonic consumer capitalist society to realize environmental sustainability and social justice.

Andrew Heffernan

The Global Politics of Local Conservation

Climate Change and Resource Governance in Namibia

palgrave
macmillan

Andrew Heffernan
Political Science
University of Ottawa
Ottawa, ON, Canada

ISSN 2731-670X ISSN 2731-6718 (electronic)
Environmental Politics and Theory
ISBN 978-3-031-24176-5 ISBN 978-3-031-24177-2 (eBook)
https://doi.org/10.1007/978-3-031-24177-2

This Palgrave Macmillan imprint is published by the registered company Springer Nature Switzerland AG.
The registered company address is: Gewerbestrasse 11, 6330 Cham, Switzerland

This book is dedicated to my grandfather Fred Collins who, although he was not able to be with us to see me through the end, by all accounts was my greatest inspiration, academically and otherwise. It is also dedicated to my dog Jagger, who also left us too soon, but who was my closest confidant, most reliable running mate, and by all accounts best friend throughout my entire adult life.

List of Acronyms

Acronym	*Meaning*
AGM	Annual General Meeting
CBNRM	Community-Based Natural Resource Management
COP	Conference of the Parties
CSO	Civil Society Organization
EU	European Union
GCF	Green Climate Fund
GDP	Gross Domestic Product
GHG	Greenhouse Gas
EFIN	Environmental Investment Fund of Namibia
INGO	International Non-Governmental Organization
IRDNC	Integrated Rural Development and Nature Conservation
MCA	Millennium Challenge Account
MET	Ministry of Environment and Tourism
NACSO	Namibian Association of CBNRM Support Organizations
NGO	Non-Governmental Organization
SWAPO	Southwest Africa People's Organization
UN	United Nations
UNDP	United Nations Development Programme
UNESCO	United Nations Economic, Scientific and Cultural Organization
UNWTO	United Nations World Tourism Organization
US	United States
USAID	United States Agency for International Development
WWF	World Wildlife Fund

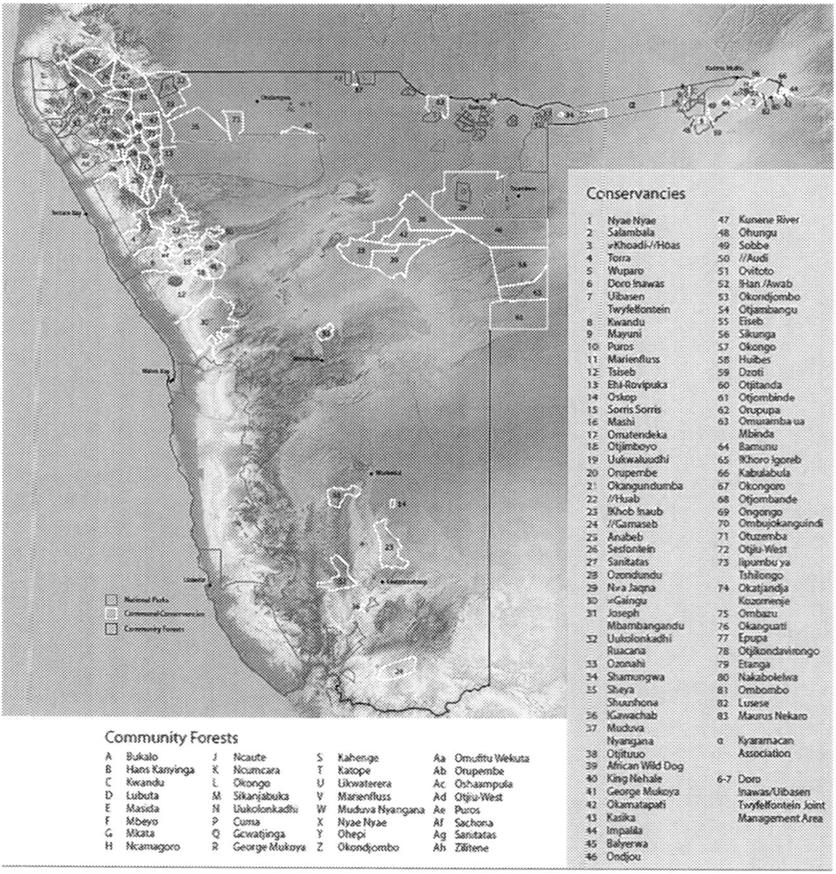

Fig. 1 Map of conservatives in Namibia (NACSO "The State of Community Conservation in Namibia" Annual Report 2020)

Acknowledgements

Rita Abrahamsen was my supervisor for my master's as well as my PhD and has consistently provided invaluable guidance, instilled an endless wealth of knowledge, and most importantly always pushed me to do the best work I possibly could. There are any number of easier pathways I could have taken along my academic journey, and when I was tempted to do so, either speaking with her directly or simply knowing what she would recommend more often than not led me to making the right—if not always the easiest—decision. Whether in the classroom, at conferences, at CIPS events, PhD workshops she would organize, or in the countless meetings we had, Rita has gone above and beyond the 'ideal supervisor' that is so often discussed abstractly. Despite being the busiest, as well as one of the most successful people I know, Rita always finds time for students, whether they be her doctoral candidates, undergraduate students, or anywhere in between—she is truly dedicated to the development of knowledge and the improvement of the academe and I aspire to be half the academic and person she is one day.

I am also endlessly grateful for my committee members Jacqueline Best and Ryan Katz-Rosene. I have had a strong and ongoing working relationship with both of them throughout the entirety of the research and writing process for this book and they have enriched my experience throughout this time as well as contributed to positive outcomes in numerous ways. Despite having extremely full plates of their own, both Jacquie and Ryan have always been extremely accessible, and have always responded positively to my endless requests, whether this be for comments on drafts, or simply input and advice on directions to take in life, they have provided me so much.

I am also deeply grateful to Chris Huggins and Larry Swatuk for the invaluable comments they have both provided throughout the process, adding to the endless workloads they both have. Their input and contributions have been essential in pushing the dissertation to the final product. Luc Turgeon was also instrumental in enabling me to put together my initial research proposal, including the literature review and theoretical framework which provided the foundations for the research that would follow.

I would be remiss if I did not offer thanks to the University of Ottawa itself. The Political Studies department has had a particular hand in this, and while the faculty and administrative support staff that have contributed to my success are far too numerous to name, I am forever indebted to you for the enriching experience that the school provides.

I will forever be indebted to kind people of Namibia, without whom this research would not have been possible. From initial contact prior to arriving, throughout my entire time on the ground, as well as through follow-up emails and calls since, the many interlocutors I was fortunate enough to work with have been endlessly warm, kind, and helpful in contributing to my many and seemingly ongoing requests. A comprehensive list of those who contributed can be found in the annex of this book and are far too many to name here. I would however like to say a special thanks to Maxi Pie Louis, Director of NACSO, who was my principal contact in Namibia and who assisted me in developing contacts throughout my journey that were pivotal to the research and to successfully navigating the vast Namibian savannah.

The work for this book has been made possible by the generous financial support of the Social Sciences and Humanities Research Council, the Ontario Graduate Scholarship, and Ontario Student Assistance Program, as well as a number of funding supports provided by the University of Ottawa. Without all of this support my research would have either been impossible or at the very least impractical due to financial constraints.

While family and friends may not always be well versed in the life of an aspiring academic, I do not know where I would be without their endless love and support. Amid the countless questions of 'when will you be done?' or 'what is that paper on again?' I have been lucky enough to have a constant support system of friends here in Ottawa, as well as family back home in Peterborough, who I was able to visit consistently throughout the years of research that have gone into this book. While it may have seemed to many like I was happy being locked in my office in Ottawa, working endless hours alone at my computer, without your love and support none of it would mean anything and I would not be where I am today.

Contents

About the Author

Andrew Heffernan holds a PhD in Political Science from the University of Ottawa where he is a part-time professor specializing in International Relations and comparative politics. His major research interests include African politics, global environmental governance, community-based conservation, and the politics of food. Andrew is also active in the Scholarship of Teaching and Learning which he is continuously publishing on, presenting at academic conferences, and implementing in his teaching in university classes. He has published in *Environmental Management, the Journal of Southern African Studies*, *Sport in Society*, *Canadian Journal of Political Science*, and *International Studies* among others. Andrew is an ongoing contributor on a project examining the Future of Sustainable Protein *(https://futureofprotein.ca/)*, as well as vice president for ISA-Canada and vice chair for the African Studies Association's Emerging Scholar Network.

List of Figures

List of Tables

CHAPTER 1

Introduction

Here people do not have the luxury of not believing in the science. Climate change is all around us, it is undeniable, and it is threatening our way of life in the immediate....[1]

Having lived in the harsh, dry conditions of the Erongo Region of Namibia for generations, Eric Xaneb and his family have always been farmers. "We are farmers," he says, "traditional farmers, historically farmers, we are farmers."[2] For Eric and other subsistence farmers in the area, life has always been challenging; the soils of the Erongo Region are not known for their productivity, and Namibia is the driest country in sub-Saharan Africa.[3] While these environmental challenges have existed for generations, the independence of Namibia in 1990 brought renewed hope particularly for the black communities of the Erongo Region. Freed from the shackles of Apartheid rule, there was optimism that land would be more fairly distributed, and all Namibians would be able to prosper from it and its natural resources. In 1996, the Nature Conservation Amendment Act

[1] Interview, Maxi Pia Louis, Director, Namibia Association of CBNRM Support Organizations, January 14, 2020.

[2] Interview, Eric Xaneb, Chairman, Tsiseb Conservancy, January 27, 2020.

[3] Jarkko Saarinen, Political Ecologies and Economies of Tourism Development in Kaokoland, North-West Namibia, Political Ecology of Tourism (Routledge, 2016).

A. Heffernan, *The Global Politics of Local Conservation*, Environmental Politics and Theory,
https://doi.org/10.1007/978-3-031-24177-2_1

was passed, which created the programme known as Community-Based Natural Resource Management (CBNRM)—a form of resource governance that devolves control of resources from central governments to local communities. Many Namibians welcomed the change as they saw an opportunity to exploit one of their key natural resources—wildlife—by managing their populations sustainably in order to benefit economically through ecotourism.[4] While CBNRM can consist of a broad set of policies and practices, Larry Swatuk summarizes it by outlining that "in theory, CBNRM is a rather elegant idea that rural livelihoods and key ecosystems in the Third World can be improved and maintained through the sustainable utilization of natural resources by states and (global) civil society."[5] Neighbouring countries such as Zimbabwe, Botswana, and Zambia were already having successes with CBNRM, and many activists, academics, governmental, and non-government organizations (NGOs) believed Namibian communities could experience similar benefits.[6]

For Eric and other farmers across the country, despite sounding attractive in theory, CBNRM also presented a number of concerns. The programme establishes conservancies which protect wildlife populations and their habitat to increase numbers, in order to encourage wealthy foreign tourists to travel to Namibia in hopes of experiencing Africa's charismatic megafauna and breathtaking landscapes. However, this shift in land use also means that any growth in CBNRM results in less new acreage available for farming, and increased wildlife numbers means greater threat to crops, livestock, and important water infrastructure. While many Namibians came to see the potential for CBNRM, and the programme was widely adopted—to the point that Namibia came to be seen as a world leader in CBNRM[7]—Eric and many of his neighbours could not ignore

[4] Julie Richardson, "Wildlife Utilization and Biodiversity Conservation in Namibia: Conflicting or Complementary Objectives?," *Biodiversity & Conservation* 7, no. 4 (1998): 549–59.

[5] Larry A. Swatuk, "From 'Project' to 'Context': Community Based Natural Resource Management in Botswana," *Global Environmental Politics*, Global Environmental Politics, 5, no. 3 (2005): 95.

[6] Fikret Berkes, Gary P. Kofinas, and F. Stuart Chapin, "Conservation, Community, and Livelihoods: Sustaining, Renewing, and Adapting Cultural Connections to the Land," in *Principles of Ecosystem Stewardship: Resilience-Based Natural Resource Management in a Changing World*, ed. Carl Folke, Gary P. Kofinas, and F. Stuart Chapin (New York, NY: Springer New York, 2009), 129–47.

[7] Dilys Roe, Fred Nelson, and Chris Sandbrook, *Community Management of Natural Resources in Africa: Impacts, Experiences and Future Directions* (London, UK: International Institute for Environment and Development, 2009).

the increased challenges as a result of this developing industry. Over time, the programme came to provide benefits for many who were directly employed through the programme, as well as in many communities broader benefits were created for members not directly employed including for farmers in the form of compensation for any losses they incurred due to wildlife. However, employment and benefits were not enjoyed equally, and while CBNRM was specifically intended to help those most marginalized in society, rarely was this the case.[8]

Eventually Eric himself became directly implicated in managing his CBNRM conservancy through a local democratic election and since that time has been a major proponent of the programme and its benefits. However, these benefits which were growing steadily and contributing to community development for years have more recently slowed. At the same time as growing wildlife populations, increased numbers of tourists and greater areas of land being protected for the conservancy programme were causing issues to farmers in some regions of the country, an entirely new and more sinister threat was posing even greater challenges—climate change. The people of Tsiseb Conservancy had long realized the climate was changing, rainfall was diminishing and becoming more erratic, and crop yields were declining. For the better part of the past decade, Namibia has been experiencing the longest and most severe drought in living memory. The drought, Eric is certain, is a result of climate change, and the type of threat that he realizes will increasingly compound existing challenges moving forward, as well as impact various groups asymmetrically. The changing climate has made farming almost impossible in a region where it had historically been challenging to begin with.[9] While for those whose livelihoods have consisted of nothing but farming, this was a painful reality to cope with, for many—especially younger generations such as Eric's daughter—the need for a shift has presented welcomed opportunities.

Eric suggests that, since independence, there has been a growing desire for modernization and for many, like his daughter, this means wanting livelihoods outside of farming. With the increasing number of constraints to subsistence farming in the area, these desires are quickly also becoming necessities. While at one time various forms of farming contributed in

[8] Arthur Frederick Hoole, "Place-Power-Prognosis: Community-Based Conservation, Partnerships, and Ecotourism Enterprises in Namibia," *International Journal of the Commons* 4, no. 1 (2009): 78–99.

[9] Interview, Eric Xaneb, Chairman, Tsiseb Conservancy, January 27, 2020.

some way to everyone's livelihood in the Erongo Region, Eric reports that as the drought stretches on, aside from mining, CBNRM-related tourism revenue is virtually all that is left for the people of Tsiseb Conservancy. The drought has left their land infertile and the incomes from tourism as well as its many offshoots have become what communities are subsisting from. Eric's position in relation to CBNRM is a complex one, and one that has similarly shifted over time. While he initially had interest in the programme as well as optimism for its success, he also realized that success could negatively impact his people's ability to generate livelihoods as they traditionally had from the land. As environmental pressures have continuously limited this ability, Eric's initial optimism about CBNRM has been renewed and he has become a fervent supporter of the programme. Eric's story connects to many others in his community and in similar communities across Southern Africa, revealing a complex set of relationships between land and livelihood, environment and economics, nature and culture, and race and gender. How we seek to understand the situation as a whole—and particularly the complex effects of climate change within it—has a lot to do with our worldview as well as our relationship with various actors who compose CBNRM programmes across Southern Africa.

While much of the early literature on CBNRM approached it as a somewhat apolitical domestic policy tool or technical fix for conservation, more recent scholarship has pointed to the complex partnerships that witness the actions of a number of local and global actors coalesce which can produce unexpected results. My research aims to build on these more recent studies, and I suggest that CBNRM cannot be adequately understood outside of global power relations.[10] While scholars have long theorized about the various forms of power that actors can wield, increasingly power has come to be understood as a phenomenon that is inherently based on the relations between actors and elements.[11] As globalization has created a world of hyper-connectivity, it becomes increasingly difficult to silo the actions of actors such as governments and policy makers. Instead, governance is more accurately viewed as being a process determined by the sometimes competing and other times harmonious actions of a plethora of

[10] Andrew Heffernan, "The Globality of Community-Based Conservation in Namibia—Centre for International Policy Studies," 2020.

[11] Eija Ranta, "State Governance and Micropractices of Power in the Process of Decolonizing the State in Bolivia," *Forum for Development Studies* 45, no. 3 (2018): 1–23.

heterogeneous actors.[12] This form of governance that has become ubiquitous in recent decades sees the efforts of governments, private interests, NGOs, interest groups, activists, and citizens coalesce in order to contribute to the ordering of society. This ordering results in layered governance apparatuses that see the above actors contribute to not only policy making, but also the implementation and execution of policies, their evolution, and norms that spread which may or may not be codified in law. Nonetheless, these norms can serve important governance mechanisms in similar or sometimes even more effective ways when accepted by a wide enough sample of a given population. One of the most effective ways to understand how governance has become layered in this way is as a global assemblage. Global assemblages serve as one analytical lens for effectively analysing increasingly complex governance arrangements that have outgrown many of the traditional theories and approaches of the social sciences.[13] CBNRM is one such governance arrangement that is enacted by a government policy, but in practice its ongoing management and evolution are much more complex than a government decree which then funnels down to the people.

Assemblages involve complex aggregate parts that can operate at once locally and globally while simultaneously affecting and being affected by their constituent parts that can be almost limitless. These parts can be synergistic but can also be conflictual and competitive. Assemblage thinking is about the multitude of actors who influence the daily transactions whether political, economic, social, or otherwise across time and space that in intricate ways constantly create and recreate power and power relations.[14] They are key to unveiling and analysing social relations that blur traditional distinctions such as social/material, subject/object, human/nature, global/local, male/female, black/white, and structure/agency. These are the numerous components that compose CBNRM as well as enable it which require a disaggregation and in-depth analysis in order to understand their interactions with one another and what these result in.

[12] Saskia Sassen and Aihwa Ong, "The Carpenter and the Bricoleur," in *Reassembling International Theory: Assemblage Thinking and International Relations*, ed. Michele Acuto and Simon Curtis (London: Palgrave Macmillan UK, 2014), 17–24.

[13] Michele Acuto and Simon Curtis, "Assemblage Thinking and International Relations," in *Reassembling International Theory: Assemblage Thinking and International Relations*, ed. Michele Acuto and Simon Curtis (London: Palgrave Macmillan UK, 2014), 1–15.

[14] Saskia Sassen, Territory, Authority, Rights: From Medieval to Global Assemblages (Princeton University Press, 2006).

Assemblage thinking is one way in which we might step outside of our usual tried, tested, and failed existing approaches. By engaging issues with many of our traditional ontological approaches 'knowing' what power states do or do not wield as a means of governing resources "might blind us to the different processes through which social and political (dis)order are assembled."[15] In order to avoid these potential issues, assemblage thinking has evolved into a tactic for developing a flattened ontology that moves away from reifying totalities and beyond the strict anthropocentrism and state-centricity of so many existing theories. This approach emerges in order to analyse complex and diverse issues without predetermined notions that seek to uncover certain types of explanations. A flattened ontology can be useful as it enables approaches to complex governance arrangements without preconceived notions of what constitutes an actor, how an actor may or may not wield power, and what types of power they might wield. With difficult issues, especially in places like Africa where a strictly defined Weberian state has likely never existed, it is this kind of humble approach that can produce some of the most accurate—though perhaps not the simplest—analyses. Developing this kind of analysis can then allow for a better understanding of the politics of CBNRM as well as the manifold actors exerting various forms of power in ways that go well beyond technical policy shifts.

A flattened ontology is likely an important part of any scholarly approach to analysing governance in the modern era; however, it is particularly salient in addressing issues in global environmental governance in order to account for questions of valuing, agency, and justice for both people and ecosystems. These are questions that are ubiquitous within issues of conservation and climate and approaching such challenges without predetermined notions of who is right or wrong, or who ought to have their voices heard, or a seat at the table vis-a-vis others is integral to achieving equitable outcomes that will be sustainable and lasting. CBNRM is one specific example of this which sees the interests of many competing actors weighed one against the other, many of whom are working to realize important goals. Often these interests overlap, but are rarely congruous and there are also competing narratives for how best goals ought to be realized. The flattened ontology is thus an important way to approach issues to account for these competing interests to attempt to realize a

[15] Abrahamsen, "Assemblages," 6.

balance of outcomes for humans, wildlife, and ecosystems—a balance that we are most often failing to achieve.

CBNRM, like many development initiatives, is highly political and it is highly complex. First, the legislation for these policies is often drafted in concert with NGOs, international institutions, as well as consultants, and each of these receives funding from a variety of actors stemming predominantly from the West. Once implemented, CBNRM programmes that result from the policies are enabled through direct support by a network of NGOs on the ground which are both domestic and international. The funding streams for these NGOs and international NGOs (INGOs) have become increasingly opaque as a variety of new development actors have emerged and traditional bilateral relationships have decreased. A parallel development has transpired with evidence suggesting that many NGOs are becoming increasingly politicized and beholden to their funders, rather than to more apolitical and altruistic goals.[16] As funding streams have narrowed, this has often led to increasingly targeted approaches to development and conservation in which rather than providing core funding for broad approaches or policies, donors—and hence the NGOs their funds are channelled through—pick certain projects, groups, communities, and individuals based on the activities they are engaged in. This book unpacks the way these shifts are leading to an erosion of democratic and transparent decision making with certain types of development flows, which is resulting in certain winners emerging based on politics, rather than evidence-based policy outcomes.

One of the major gaps in the existing literature is the failure to adequately account for the impacts of climate change on CBNRM as will be demonstrated in Chap. 2. This omission occurs as CBNRM is often framed as an issue of domestic conservation that is separate from the broader issues of global climate change. However, climate change is posing major challenges to CBNRM programme success and as such ought to be central to any analysis of it in order for policies to evolve to account for these shifting realities. Therefore, this book seeks to answer the following research questions: (1) How has CBNRM evolved to respond to the challenges of climate change? (2) To what extent do changes and adaptations benefit local people?

[16] Stefan Toepler et al., "The Changing Space for NGOs: Civil Society in Authoritarian and Hybrid Regimes," *Voluntas (Manchester, England)* 31, no. 4 (2020): 649–62.

In order to answer these questions, it is necessary to understand the complexities of CBNRM and its place within global power relations. Drawing on recent literature on global assemblages,[17] I analyse CBNRM as a policy which results in a complex layering of governance that both enables and requires a multitude of heterogeneous actors to contribute to the implementation of certain types of policies. As a result, although the state remains one of my components of inquiry, it is not central to my analysis and instead plays varying roles, in diverse ways at various times and in a sense may seem to come and go as the argument and analysis unfold. Instead, what is central to the present inquiry is the actors which constitute CBNRM and the relations between them as well as the ways in which these are central to environmental governance and environmental politics in Africa. This will lead me to my central argument, that the local/global nature of CBNRM enables it to adapt to both exogenous and endogenous pressures and challenges. However, due to power imbalances programme evolution in many ways sees a continuation of business as usual, which means that often communities are not empowered and do not benefit in the way the policy entails. Furthermore, although the predominant literature on CBNRM has omitted climate change from ongoing debates, those on the ground are actively engaged in adapting to and mitigating its harmful effects. While climate change poses some of the greatest challenges to CBNRM, it is also providing certain opportunities that are enabling the programmes to take part in a necessary evolution in order to remain relevant and continue garnering success in achieving sustainable development goals. To be sure, I am not arguing that climate change provides net benefits to the programme, but opportunities in this context do not necessarily connote something intrinsically positive. Instead, I analyse opportunities that indicate how adaptation and necessary responses to climate change provide avenues for some actors to adopt new strategies for profitable enterprises and survival. While the CBNRM programme is evolving in important ways in response to climate change and becoming more sustainable, climate change has also resulted in a proliferation of local and global partnerships which are resulting in tensions that have led to continued challenges in attaining the degree of community empowerment that

[17] Rita Abrahamsen and Michael C. Williams, *Security beyond the State: Private Security in International Politics* (Cambridge, UK; New York: Cambridge University Press, 2011); Acuto and Curtis, "Assemblage Thinking and International Relations"; Glenn C. Savage, "What Is Policy Assemblage?," *Territory, Politics, Governance* 8, no. 3 (2019): 1–17.

CBNRM initially promised. As such, there are ways that CBNRM might better be referred to as 'International' Community-Based Natural Resource *Governance*, as it is best understood as an iteration of global environmental governance, rather than a domestic policy for conservation. While some existing work points to important global dynamics, this book will employ an assemblage approach to demonstrate the ways in which this complex assemblage evolves in response to the impacts of climate change in novel ways.

Another central point the book will analyse centres around questions of gender, race, and the marginalized populations in local communities that CBNRM was intended to benefit. Following independence, the shift in land tenure and move to CBNRM encouraged black Namibian and particularly women, that they would have a renewed chance at improved outcomes towards equality. However, equality has largely been failed to be gained. Nonetheless, recent years have seen a sort of changing of the guard with an increase in women in positions of power within conservancies and NGOs as well as an increase in black Namibians taking over these same roles. Traditionally, particularly for the country's main CBNRM NGOs these roles have been filled by white people, and often people that stem from outside of Namibia. There has been an effort to remedy these ongoing issues in order for Namibians to not only engage in participation, but also realize a true ownership over their own development. These challenges are linked with climate change which is exacerbating them while also providing opportunities for renewal of the programme and those who compose it in the country.

My main contributions in this book are to inform both academic and practitioner communities in order to mobilize knowledge on the increasing impacts of climate change on CBNRM, as well as to analyse the complex network of actors that impact the evolution of this form of resource governance in response to these proliferating challenges. This will enable scholars to build on the existing literature to continue to better understand CBNRM and the ways in which it rearticulates power between actors from the local to the global levels. It will also provide empirical results that practitioners can leverage in order for them to continue to make informed decisions with practical policy relevance in light of developing challenges and opportunities. While a plethora of development studies scholars have forwarded a wealth of literature on analysing the strengths and weaknesses of CBNRM as a means for conservation and sustainable development, as well as many having looked specifically at the impact of global organizations on CBNRM, many of these remain too locked within specific confines, paying less attention to power and politics and to the wealth of heterogeneous actors that not only impact CBNRM, but actually led to its very genesis, as well as contribute to its ongoing management and evolution which is key to the book.

In order to analyse this multitude of actors and their social relations I employ a global assemblages approach to analyse CBNRM with a flattened ontology, which means I first and foremost do not engage with the subject assuming it is a policy that is passed by the Namibian government and then flows down to its subjects as is often traditionally understood of many policies. Instead the flattened ontology that is part of the assemblage approach sees me trace the developments of this policy through historical analysis, and then builds through to discourse analysis of secondary sources, grey literature, and interviews in order to piece together the heterogeneous and often disparate actors whose efforts and interests combined in cooperative and often simultaneously competitive relations to result in the unofficial beginning of CBNRM prior to independence. CBRNM in Namibia was eventually codified in the 1996 Nature Conservation Amendment Act and over the past 25 years has continued to evolve in response to challenges and events. But the legislation does not keep evolving in the halls of government and instead this evolution is a result of the social relations of the actors that are traced in the book. This study analyses the power relations that emerge from this governance assemblage and builds on Foucauldian conceptualizations of power to analyse productive power as elaborated by Barnett and Duval in order to unpack ways in which various actors are able to achieve their ends.[18] I suggest that CBNRM is not a domestic policy for conservation that is then influenced by global power relations, but instead it is a policy assemblage that emerges through a confluence of global and local, public and private social relations and that it cannot be understood outside of these. Below I will further unpack some of the complexities of CBNRM as a form of resource governance in order to contextualize the chapters to come.

[18] Michael Barnett and Raymond Duvall, *Power in Global Governance* (Cambridge, United Kingdom: Cambridge University Press, 2004).

The Local/Global Nature of Wildlife Governance

A key example of the complexities of the interconnections that both produce and enable CBNRM is with the various relationships heterogeneous actors have with wildlife. The charismatic megafaunas of Southern Africa are some of the most majestic and alluring species on earth. As children, many play with toy elephants, lions, rhinos, giraffes, and cheetahs, while these same animals occupy popular TV shows and movies. Africa becomes a place of great intrigue to children in much of the world as the unknown, far off, almost mythical home of these great beasts. As adults, these beautiful creatures often adorn walls as art and are consumed through documentaries extolling their wisdom, serenity, and power. Many dream of perhaps one day travelling to the continent where so many of the earth's unique and rare wildlife reside. Those lucky enough to possess the means to do so might even travel thousands of kilometres to submerse themselves in nature and realize their dream of going on safari to experience these rare animals first-hand.

While for the vast majority of the world's population—especially in the West where most safari goers stem from—seeing Africa's wildlife first-hand will never become a reality, virtually everyone would agree that protecting them is a desirable end. Beyond the beauty of the wildlife themselves and the travesty that would be losing such species to extinction, their survival is inherently linked to the maintenance of healthy and functioning ecosystems. Among the World Wildlife Fund's (WWF) list of endangered, vulnerable, and threatened species are those named above, as well as many more that both we and our children can most likely easily name.[19] Unfortunately, it is those which most capture the human imagination and wonder that are similarly facing the greatest threat. This is no accident, as by now it will come as no surprise that it is humans and human activities that have directly caused the gradual decline of wildlife populations around the globe, beginning with the colonial era and the Industrial Revolution.

Since the beginning of the twentieth century, most of Africa's charismatic megafauna have been experiencing drastic declines brought on by a variety of human activities. This has included the destruction of habitat for the construction of settlements, farming activities, and use for fuel.[20] It has

[19] WWF, "Species List | Endangered, Vulnerable, and Threatened Animals | WWF," World Wildlife Fund, n.d.

[20] David Hulme and Marshall Murphree, *African Wildlife and Livelihoods : The Promise and Performance of Community Conservation* (Great Britain: Woolnough, Irthlingborough, 2001).

also tragically been exacerbated by the slaughter of huge numbers of animals for parts such as ivory, rhino horns, furs, and skins to be sold on the black market. Initially many of these items were desired across the colonial world as luxury and decorative goods. Eventually this black market also proliferated in Yemen as a result of the influx of capital from neighbouring Saudi Arabia in the 1970s which led to a substantial—and for rhinos deadly—increase in demand for djambias (daggers worn as a status symbol by wealthy Yemeni men) made from rhino horn.[21] Today, the traffic of these illicit goods continues largely in Asian markets as people continue to falsely believe in their medicinal properties.[22]

A further and growing threat to wildlife is climate change. Scientists have warned that we are on the brink of a sixth mass extinction, and it is clear that this is already well underway.[23] Many regions of Africa are experiencing greater effects from climate change than the global average and this is drastically altering precipitation and migration patterns while forever altering ecosystems that are key to sustaining healthy flora and fauna.[24] Thus, while the challenge of conserving wildlife is often framed as a depoliticized issue in that 'we all agree they must be saved,' the issue remains highly political and there is much less agreement on *how* this ought to be accomplished as well as importantly *who* ought to be responsible for doing so.

In response to the negative impacts of climate change, the CBNRM programme has begun to evolve. While certain actors that were key components of the programme during the early years of CBNRM have dissipated, others have joined with a renewed fervour. Actors like the Green Climate Fund (GCF) have partnered with the Environmental Investment Fund of Namibia (EFIN) and various conservancies in order to procure global funds to help make the programme more sustainable. In some cases, this has resulted in the building of solar power plants for high-end

[21] Garth Owen-Smith, *An Arid Eden*, Illustrated edition (Johannesburg: Ball Jonathan Pub, 2011), 392.

[22] Alfan A. Rija et al., "Global Extent and Drivers of Mammal Population Declines in Protected Areas under Illegal Hunting Pressure," *PloS One* 15, no. 8 (2020): e0227163–e0227163.

[23] Gerardo Ceballos, Paul R. Ehrlich, and Peter H. Raven, "Vertebrates on the Brink as Indicators of Biological Annihilation and the Sixth Mass Extinction," *Proceedings of the National Academy of Sciences* 117, no. 24 (June 16, 2020): 13596–602.

[24] Walter Leal Filho, *Experiences of Climate Change Adaptation in Africa*, Climate Change Management (Springer Berlin Heidelberg, 2011).

lodges, the retrofitting of buildings to make them easier to keep cool, more efficiently providing freshwater, or the harnessing of greywater. These activities demonstrate innovative ways in which the international community continues to partner with local communities in order to ensure that CBNRM can adapt to and mitigate the harmful effects of climate change. These can be positively pointed to as benefits derived from novel partnerships that have emerged in response to climate change. However, these partnerships are highly political and can lead to an increased foothold for international actors that are increasingly divorced from democratic decision making. The resultant social interactions often result in power imbalances that must be carefully analysed in order to understand how decision making takes place and whose interests are being realized.

The question of how to protect biodiversity in general, and charismatic megafauna in particular, is almost as old as the problem itself. Even in the early days of imperialism and European encroachment onto the African continent, many colonial administrators understood the beauty of these beasts and took official actions in order to ensure their preservation. While these same colonial regimes provided the catalyst for the major decline in wildlife populations over the twentieth century, many also began developing various approaches to conservation. Today, this legacy remains largely in the form of many of Africa's major national parks which provide refuge for species whose natural habitat continues to disappear at alarming rates in some places.[25]

This type of approach that walls off and limits human use and at times even interaction with wildlife has come to be known as fortress conversation. For many, this offers the most effective solution to protect wildlife. By limiting the ability of local human populations to encroach on habitat and by offering a strong layer of protection against poaching, it is argued wildlife will be secured.[26] In theory, this approach is sound, and in practice it has provided some measure of success. However, despite decades of increased efforts at various forms of fortress conservation, wildlife populations continued to dwindle across the continent and this tragedy has been paralleled by protracted underdevelopment for local human populations,

[25] Mirja Stoldt et al., "Transfrontier Conservation Areas and Human-Wildlife Conflict: The Case of the Namibian Component of the Kavango-Zambezi (KAZA) TFCA," *Scientific Reports* 10, no. 1 (May 14, 2020): 7964.

[26] David Hulme and Marshall Murphree, "Communities, Wildlife and the 'New Conservation' in Africa," *Journal of International Development* 11, no. 2 (1999): 277–85.

and predominantly black populations across Africa.[27] In response to these failures, over time, the predominant thinking in academic circles as well as development discourse came to focus on harmonizing the sustainable development of human populations alongside local natural resources including wildlife.

These issues are by no means limited to wildlife populations, and there are a number of other natural resources which are simultaneously key to sustainable development for human populations as well as to ecological health and well-being, such as freshwater resources, and forests and forest products. In order to treat the dual issue of unsustainable resource extraction mixed with protracted underdevelopment that has been experienced under the fortress conservation paradigm, CBNRM has developed across much of Southern Africa as a key resource governance technique. CBNRM consists of a broad set of policies and practices and can take a variety of distinct forms. Such differences aside, it generally involves the devolution of control of certain natural resources from national governments into the care and control of local communities. The idea being that those who reside close to resources will manage them more sustainably as the necessity for conservation for their own future will be more clearly linked and they will also be held to account by family, friends, and neighbours, rather than by distant governments.[28] In Africa it has not been unusual for relations between national governments and rural communities to be fraught and the state is often viewed as a purveyor of insecurity rather than security, and a plunderer of resources, rather than a provider of services.[29] This is in part why decades of fortress conversation witnessed control over natural resources centralized under the purview of the state, while limiting any meaningful input or decision making from those residing in close proximity to the resources.

CBNRM developed as a response to fortress conservation which came to be perceived as a neocolonial approach to ecological preservation. While CBNRM evolved as a form of global environmental governance broadly, it also did so in highly localized and particular ways in various

[27] Amity A. Doolittle, "Fortress Conservation," in *Encyclopedia of Environment and Society*, 5 vols. (Thousand Oaks: SAGE Publications, Inc., 2007), 705–705.

[28] Russell Taylor, "Community Based Natural Resource Management in Zimbabwe: The Experience of CAMPFIRE," *Biodiversity and Conservation* 18, no. 10 (2009): 2563–83.

[29] Michael Bratton and Nicholas van de Walle, "Chapter 2: Neopatrimonial Rule in Africa," in *Democratic Experiments in Africa Regime Transitions in Comparative Perspective* (Cambridge University Press, 1997).

settings.[30] CBNRM is intended to devolve power from the state to local communities.[31] This form of resource governance has a long history and evolved over time before growing in popularity in the 1980s and 1990s when dominant thinking in development and global environmental governance came to focus on community empowerment, partnership, and ownership.[32] This was a time of a marked shift in which neoliberalism can be said to have won out across much of the world as the dominant ideology.

For CBNRM, Western involvement entailed the allocation of donor funding through a proliferating network of largely white and male-led NGOs as well as private interests across Southern Africa to invest in projects that were intended to promote sustainable development. In many cases this involved devolving the rights to certain natural resources—such as wildlife—to local populations who would become responsible for both the costs associated with managing resources and possible benefits that could be accrued from doing so responsibly and sustainably. Benefits were generally derived from the commoditization of resources through a number of industries based on various forms of ecotourism.[33] Often local communities would engage in joint-venture agreements with private tourism companies who would build lodges, campsites, infrastructure for safaris, as well as the vast economies that support tourism industries including restaurants, guided hikes and tours, national and world heritage sites, local craft and ware shops, museums, beaches, and more. While many of the more prestigious and well-paying jobs were occupied by those already most well off in society, many of these economic offshoots also provided avenues for some of society's most marginalized communities to develop livelihoods. This development was based on wealthy tourists travelling to experience the African savannah and its rare and majestic wildlife.[34] One of

[30] Owen-Smith, *An Arid Eden.*

[31] Brian Jones and Marshall Murphree, "The Evolution of Policy on Community Conservation in Namibia and Zimbabwe," *African Wildlife and Livelihoods: The Promise and Performance of Community Conservation*, 2001, 38–58.

[32] Martin Sjöstedt, "Aid Effectiveness and the Paris Declaration: A Mismatch between Ownership and Results-Based Management?," *Public Administration and Development* 33, no. 2 (2013): 143–55.

[33] Sheona Shackleton and Bruce Campbell, Devolution in Natural Resource Management: Institutional Arrangements and Power Shifts: A Synthesis of Case Studies from Southern Africa (CIFOR, 2001).

[34] C. Fabricius and S. Collins, "Community-Based Natural Resource Management: Governing the Commons," *Water Policy* 9, no. S2 (November 2007): 83–97.

the key factors for success is that the savannah must remain natural, and the wildlife must remain healthy and abundant. Thus, the more successfully any country or community protected its wildlife and their habitat, the more economic development they could enjoy—ideally resulting in an intended win-win scenario. This win-win scenario is precisely what Elinor Ostrom based her seminal work *Governing the Commons* on, as a way to prevent collective resource issues from degenerating into tragedies of the commons.[35] The 'tragedy of the commons' is the term first used by ecologist Garret Hardin to describe what he suggested was an inevitable collapse in common-pool resources due to competitive overuse and extraction, or market failures in economics terms.[36] Managing resources in order to profit from them in long term is generally seen as sustainable and is intended to be both broad-based and equitable as jobs can be created and incomes can be reinvested in communities through infrastructure projects, education, as well as cash and in-kind benefits that can be distributed to community members.

This form of sustainable development is also intended to consist of a third important aspect beyond environmental conservation and economic development, which is community empowerment—creating a triumvirate of goals. These goals in turn importantly form the three pillars of CBNRM.[37] CBNRM is intended to devolve power to communities so they can develop local democratic governance mechanisms. This institutional development is intended to provide communities and their members a voice in their own development in order to achieve the 'ownership' and 'participation' on which CBNRM was based.[38] Many scholars have pointed to this third pillar of CBNRM as the most important, as ownership and participation are intended to engender long-term sustainable

[35] Elinor Ostrom, *Governing the Commons: The Evolution of Institutions for Collective Action*, The Political Economy of Institutions and Decisions (Cambridge University Press, 1990).

[36] Garrett Hardin, "The Tragedy of the Commons," *Science* 162, no. 3859 (1968): 1243–48.

[37] Andrew Heffernan, "Development, Conservation, Empowerment: The Trilemma of Community-Based Natural Resource Management in Namibia," *Environmental Management*, 2022.

[38] Louise Fortmann, Emery Roe, and Michel van Eeten, "At the Threshold between Governance and Management: Community-based Natural Resource Management in Southern Africa," *Public Administration and Development* 21, no. 2 (2001): 171–85.

development that evolves beyond resource management.[39] Despite the importance of community empowerment, these are the least well-explored dynamics within the existing literature.[40] This is due in part to it being the most abstract and difficult to quantify of the three pillars of CBNRM, while simultaneously the hardest to achieve.

These are some of the complexities this book will analyse including the ways in which the CBNRM programme is adapting and evolving in response to the impacts of climate change, as well as how various actors are driving such responses. It will place a specific focus on the politics involved in the complex governance arrangements that emerge as well as analyse to what degree communities benefit and are empowered and those most marginalized within communities are impacted. The next section outlines how the book will unfold.

Chapter Overview

To set the scene for CBNRM as a form of global governance Chap. 2 begins by discussing a brief history of the shifts that occurred in the global political economy, highlighted by the complexification of society and limitations faced by the state. The chapter also discusses the end of bi-polarity and the resultant effects on the international order. These changes are key to understanding the shift from fortress conservation towards community-based forms of conservation. These shifts occurred not only in order to empower communities, but also as a response to a broad realization that in many cases the state could not effectively govern every aspect of society. As a result, various techniques of governing at a distance emerged and this included efforts to rebound society's challenges back onto it, in order for society to become implicated directly in developing solutions. Part of what this has resulted in is shifts in the type of power that is exercised as well as the types of actors involved in governance. Thus, Chap. 2 will next turn to analysing power and its polymorphous character. This includes conceptualizations of power beyond a traditional understanding of coercive power that is wielded by sovereign decree. Increasingly scholars have posited that

[39] Brian Child, "The Practice and Principles of Community-Based Wildlife Management in Zimbabwe: The CAMPFIRE Programme," *Biodiversity & Conservation* 5, no. 3 (1996): 369–98.

[40] Karol Boudreaux, "A New Call of the Wild: Community-Based Natural Resource Management in Namibia Essay," *Georgetown International Environmental Law Review* 20 (2008): 297–336.

power is more diffuse and produced by a variety of actors based on their social interactions. Following this brief macro-level analysis, the chapter looks more closely at the literature on CBNRM specifically that unpacks what this form of resource governance is, what its goals are, and how it is intended to accomplish these. It also briefly explores the history of CBNRM. The chapter also discusses why existing approaches have yet to explain the evolution of CBNRM in response to climate change. Ultimately, it will outline key gaps in the literature which include analyses of climate change and the evolution of the programme in order to adapt to these realities as well as the degree to which these evolutions serve to benefit local communities.

In Chap. 3, I outline my theoretical framework, drawing on recent International Relations literature on global assemblages. I provide a brief overview of the more philosophical foundations of assemblage thinking, prior to discussing some of the ways it has evolved, specifically within the discipline of IR as well as ways it has been employed to understand various complex policy formations in the social world. The chapter goes on to discuss the ways scholars have developed global assemblages beyond a traditional theoretical framework and ontology and to also serve as a methodological approach for unpacking and analysing complex phenomenon. This methodology section will include an elaboration of Namibia as my case study, my approach to fieldwork, as well as process of data analysis and triangulation. Namibia was chosen for this study as the crucial case design as it is recognized as a global leader in CBNRM due to having devoted the highest percentage of land to this form of resource governance on earth, as well as having a recently drafted and progressive constitution with stipulations for environmental conservation written throughout its sections.

Next the book turns to the empirical chapters with Chap. 4 outlining the genesis of CBNRM in Namibia which will include a brief history from the colonial period, through to the Apartheid era, and onto independence in 1990. This history will include key relevant background on the politics of the country including its de facto one-party state system that has emerged since independence. It will discuss how, why, and as a result of whose actions CBNRM developed in the country, how it grew in popularity, as well as what form it took on and what its main activities consisted of. This will incorporate a discussion of conservation activities occurring in the country prior to independence with a number of activists, academics, and development workers on the ground. These actors coalesced through the changes brought on by independence and the parallel shift in the international system in 1990 which afforded them a wealth of new opportunities. These new opportunities led to the formation of a network of NGOs that worked with consultants and a broad array of international donors in order to not only draft Namibia's CBNRM legislation, but also implement it with their own influence in various regions across the country. Through the genesis of CBNRM, this chapter begins unpacking the ways in which this form of resource governance is about much more than a straightforward devolution of state power to local communities. Beyond this, the chapter discusses questions of local-level corruption which are key to understanding the degree to which communities benefit from CBNRM. This form of resource governance was intended to circumvent the type of neopatrimonial practices and large-scale corruption that have often been seen at the state level in Africa.

In Chap. 5, I outline the impacts of climate change on Namibia in general, and on the CBNRM programme in particular. This begins with discussion of the changing climate and how it affects livelihoods, tourism, wildlife, and their natural habitat and the many ways climate change is exacerbating almost every previously existing challenge to CBNRM in Namibia while also outlining the ways this has largely been ignored, or at least not been connected to CBNRM and programme outcomes. Namibia has been experiencing a severe and protracted drought for the better part of a decade and while droughts are a normal part of the hydrological cycle, their severity, duration, and frequency are all increasing which is threatening Namibians as well as the flora and fauna across the country. Decreasing precipitation patterns are making life more challenging for everyone and have begun to severely impact the ability of the CBNRM programme to produce benefits that had been enjoyed in the first 15 years of its implementation. The chapter will discuss widespread losses in livestock, veld food for domesticated livestock as well as wildlife, and the increasing pressure this all places on humans and wildlife living in close proximity—this being the essence of CBNRM. The chapter then turns to discuss opportunities that are being provided by some of the very same challenges discussed above. Despite its absence from much of the literature, the chapter demonstrates that people on the ground are acutely aware of the impacts of climate change and are active in efforts to both adapt to and mitigate further harmful effects. This has resulted in novel partnerships which have witnessed new actors amalgamate into the programme as others have dissipated over time. New global actors have joined and are channelling funds to make various parts of the programme more sustainable; however, the question remains who these funds are being directed to benefit and whether those the programmes espoused to help are left better off at all. Many of these actors such as the Green Climate Fund are both new partners to the programme and new *types* of actors that have emerged on the global stage. They are being empowered with increasing legitimacy by the international community and being supported with ever more funding and decision-making autonomy regarding development activities which is having important effects on how and where funding is allocated. In many cases this is leading to a disembedding of development from democracy which can mean funding goes to those who are connected to powerful state and global actors, rather than where it is needed most or might contribute to the greatest results.

In the penultimate chapter, I discuss the ways in which CBNRM is able to adapt to a variety of forces and who benefits as a result of these evolutions. Chapter 6 unpacks the actors that are key in driving the evolutions which are integral to understanding why, how, and in reaction to what the programmes adapt. Part of this adaptation is importantly with the make up of the programme itself in the sense of who holds positions of power and how these positions relate to questions of gender and race. In order to understand this, we must analyse power imbalances that exist between a variety of actors and specifically between local communities and more powerful global actors who are often able to realize their interests. These powerful global actors often have interests that shift over time in response to changing global fashions regarding interests in specific types of conservation, development, and even tourist destinations. These include new ecotourism destinations which shift over time, the growing aversion to trophy hunting as an approach for conservation income, as well as the shift from national parks to more integrated conservation approaches. These shifting fashions that emerge within the international community end up having substantial impacts on the ground regarding what types of activities are favoured and hence funded. It becomes clear that going local is not sufficient in itself to prevent corruption, and as a result, politics and power relations must be analysed from both global and local levels.

Finally, Chap. 7 outlines conclusions regarding the evolution of CBNRM in response to climate change. While it is clear that climate change has presented a number of challenges to the CBNRM programme, it has also presented a number of opportunities which are being harnessed by certain actors which among other benefits is helping to diversify those holding positions of power. This is presenting renewed avenues for actors to join CBNRM governance in order to pursue their own interests, while at times also contributing to congruous goals of those who have long been involved in CBNRM. Despite this, goals are often divergent and have also continued to contribute to ongoing tensions as actors vie to realize their aims and objections. Beyond ecological concerns, major questions persist about the degree to which local communities are empowered, and beyond that the degree to which those most marginalized in these communities are provided any kind of voice or are at all better off. It is clear that CBNRM produces some benefits—specifically wildlife conservation and a degree of economic development—and that it has and continues to realize important and demonstrable goals. However, equally clear is that it does not accomplish all the goals its proponents suggest—specifically community empowerment and broad-based equitable economic growth—and that these issues are in some senses being further exacerbated by the impacts of the changing climate. Climate change is challenging the viability of the programme itself while also creating avenues for new types of actors such as the Green Climate Fund who are afforded legitimacy and funding by the international community, but who do not function in the same way as traditional multilateral organizations. This is enabling avenues for these novel hybrid actors to exert their power and influence in development contexts to pick winners based on their own criteria which in some ways is continuing the same type of approaches to development that have been tried, tested, and failed. Thus, long after communities were supposed to have become self-sufficient, what we are seeing is a proliferation of actors involved, and a failure for communities to attain autonomous self-governance. As a result, CBNRM might best be described as International Community-Based Natural Resource Governance in which a complex assemblage governs resources at the local level. While in some ways, ground continues to be gained on this path, and in others, this autonomy seems to be a distant goal that will remain unachieved.

CHAPTER 2

Contemporary Evolutions in Global Environmental Governance

As a result, various techniques of governing at a distance emerged and this included efforts to rebound society's challenges back onto it, in order for society to become implicated directly in developing solutions. Part of what this has resulted in is shifts in the type of power that is exercised as well as the types of actors involved in governance. This includes conceptualizations of power beyond a traditional understanding of coercive power that is wielded by sovereign decree. Increasingly scholars have posited that power is more diffuse and produced by a variety of actors based on their social interactions. Following this brief macro-level analysis, the chapter looks more closely at the literature on CBNRM specifically that unpacks what this form of resource governance is, what its goals are, and how it is intended to accomplish these. It also briefly explores the history of CBNRM. The chapter also discusses why existing approaches have yet to explain the evolution of CBNRM in response to climate change. Ultimately, it will outline key gaps in the literature which include analyses of climate change and the evolution of the programme in order to adapt to these realities as well as the degree to which these evolutions serve to benefit local communities.

A. Heffernan, *The Global Politics of Local Conservation*, Environmental Politics and Theory,
https://doi.org/10.1007/978-3-031-24177-2_2

> *In terms of governance, the world is too disaggregated for grand logics that postulate a measure of global coherence.*[1]

While much of the literature on CBNRM approaches it as a domestic policy tool for conservation, I place it firmly within transformations in global environmental governance. This chapter provides a conceptual and theoretical understanding of CBNRM from its emergence within global environmental governance and development discourse, through to its implementation and evolution over several decades. It starts by discussing some of the key shifts within global politics that have been reflected in the literature on global environmental governance. This includes a shift away from govern*ment* to increasing approaches to gover*nance*. These shifts in governance mean that increasingly governments are outsourcing or devolving functions traditionally administered directly by the state apparatus. Resultantly, beginning in the 1970s and accelerating through the 1980s and 1990s there has been a push by many within government and without to decrease coercive approaches to exercising power by extending rights to govern a number of aspects of life. The overt impetus for doing so is to responsibilize various groups and communities so that they are directly implicated in developing solutions to issues they face. Beyond simply shifting responsibilities, however, these evolutions in governance have led to a reconfiguration of the way power and power relations manifest. This includes who wields power, what types of actors are able to do so effectively, as well as a proliferation of the types of power that are produced beyond traditional conceptions of coercive power. As a result, scholars have increasingly analysed various forms of power including sovereign, disciplinary, and productive power.[2]

Situating CBNRM within the literature on global environmental governance and unpacking these forms of power is integral for analysing the way the programmes are operationalized, as well as the ways in which they are able to adapt and evolve in response to the changing world around them. CBNRM must be understood as more than an apolitical domestic technical policy tool for environmental conservation, and instead is best understood as a form of resource governance that is inextricably connected to a multitude of global actors. It is also intimately connected to

[1] James N. Rosenau, "Governance in the Twenty-First Century," *Global Governance* 1, no. 1 (1995): 16.

[2] Michel Foucault, "The Subject and Power," *Critical Inquiry* 8, no. 4 (1982): 777–95.

the international community as its ethos is centred around the protection of endangered species that much of the world feels are a global good which must be protected at all costs. As a result, a plethora of global actors involved in CBNRM activities on the ground are afforded a specific degree of power and legitimacy based on global populations who see their efforts to protect wildlife as an unquestioned good in many respects. Beyond these conservation aspects, CBNRM is specifically intended to engender sustainable development for local communities, and as a result their interests are inherently tied up in programme success as well. All of this creates a complex phenomenon that has been produced in part as a result of the evolutions in global environmental governance which are discussed in this chapter.

Transformations in global politics have led to increased debates on, and interest in, global governance as a concept that quickly developed following the end of the bipolar era of the Cold War. The 1995 UN Commission on Global Governance defines governance as including the formal institutions and regimes empowered to enforce compliance, as well as the informal arrangements that people or institutions have agreed on or perceive to be in their interest.[3] Global governance includes traditional types of bureaucratic government but also goes far beyond these in an attempt to account for the broad networks and social interactions made possible by the forces of globalization. These networks lead to both the need for and the possibility of governance outside traditional state means. Rather than simply shifting from a bipolar order to a post-war unipolar system, global governance assumes that a more multicentric and globalized world emerged at the beginning of the 1990s ushering in a similarly complex and dynamic new global politics.[4] Margaret Keck and Kathryn Sikkink discuss these evolutions as the emergence of transnational networks in an excerpt that is worth quoting at some length here:

> World politics at the end of the twentieth century involves, alongside states, many nonstate actors that interact with each other, with states, and with international organizations. These interactions are structured in terms of networks, and transnational networks are increasingly visible in international

[3] Commission on Global Governance, *Our Global Neighbourhood: The Report of the Commission on Global Governance* (Oxford; New York: Oxford Paperbacks, 1995).

[4] Rosaleen Duffy, "Global Environmental Governance and the Challenge of Shadow States: The Impact of Illicit Sapphire Mining in Madagascar," *Development and Change* 36, no. 5 (2005): 825–43.

> politics. Some involve economic actors and firms. Some are networks of scientists and experts whose professional ties and shared causal ideas underpin their efforts to influence policy. Others are networks of activists, distinguishable largely by the centrality of principled ideas or values in motivating their formation.[5]

In particular, this has demonstrated the inadequacy of traditional frameworks that place utmost emphasis on state to state relations based on territoriality and instead on the growing importance of complex networks of actors involved in global politics that are simultaneously public and private, global and local.[6] The increasing interest in global governance can be seen as a shift towards the acceptance that with any issue in global politics, there is a multitude of actors and networks that operate in intricate and novel ways to exercise various forms of power and influence outcomes. Examples of this include the proliferation of global multilateral institutions, the rise of multinational corporations, the ever-growing importance of global financial flows, and even the rise of terrorist groups such as ISIS, Boko Haram, and Al-Shabaab.

This multiplication of actors in global governance has led to what some deem public-private governance networks which are

> composed of public actors, private firms, interest organizations (e.g. user group associations, small-scale farmers, trade associations, etc.). In this arrangement policies do not emerge from concerted and programmed actions at government level, but rather from interactions between these actors.[7]

These types of public-private governance networks have become ubiquitous and formulate an important part of how global politics now functions at all levels. As a result, scholars have argued that the state is experiencing an erosion of sovereignty.[8] Many conceptualize this as an erosion in the

[5] Margaret E. Keck, *Activists beyond Borders: Advocacy Networks in International Politics* (Ithaca, N.Y.: Cornell University Press, 1998), 1.

[6] James Rosenau, "Governance in the Twenty-First Century," *Global Governance* 1, no. 1 (January 1, 1995): 13–43.

[7] Albertina Ndeinoma, K. Freerk Wiersum, and Bas Arts, "The Governance of Indigenous Natural Products in Namibia: A Policy Network Analysis," *Environmental Management* 62, no. 1 (July 1, 2018): 31.

[8] Siba N. Grovogui, "Regimes of Sovereignty: International Morality and the African Condition," *European Journal of International Relations* 8, no. 3 (2002): 315–38.

state's capacity to exercise authority over its own interests—at least in the way it was once able to.[9] In an early account of these developments, Hall and Biersteker posit "this confronts national governments with a trade-off between efficiency and autonomy, and it appears that most national governments are opting for the former."[10] They explain that power and authority are closely related, but while power enables one to exercise one's will and achieve ends, authority embodies a degree of legitimacy.

Global environmental governance is a specific issue area in which we can very clearly see this multiplicity of actors and how the resultant disaggregation of power plays out. Rosaleen Duffy argues that "the increasing interest in global governance within debates about the nature of international relations, development, and environmental management can be regarded as part of a fundamental shift away from thinking about states as the most important political actors in the global system."[11] From this, the environment has proven to be one of the most compelling examples of these evolutions in global politics as it is an issue that knows no territorial bounds as well as being one that has proven to not be effectively treated at the state level alone. Instead, environmental governance is enacted through a complex network of institutions and authorities at all levels. Examples of this include endangered wildlife populations, which are governed through the Convention on the International Trade in Endangered Species, as well as forests, which fall under the International Timber Trade Council. Another useful example for understanding the way environmental governance is often operationalized either outside of or in the absence of the state is with the 2017 declaration by then President Donald Trump that the US would pull out of the Paris Agreement alleging it had 'damaged' the US economy.[12] Despite this reversal of national policy, as well as the announcement of the 'America First Energy Plan' that would see an

[9] Jens Bartelson, "Three Concepts of Globalization," *International Sociology* 15, no. 2 (2000): 180–96; Jacqueline Best, "Why the Economy Is Often the Exception to Politics as Usual," *Theory* 24, no. 4 (2007): 87–109; Stuart Elden, "Missing the Point: Globalization, Deterritorialization and the Space of the World," *Transactions of the Institute of British Geographers* 30, no. 1 (2005): 8–19.

[10] Rodney Bruce Hall and Thomas J. Biersteker, eds., *The Emergence of Private Authority in Global Governance* (Cambridge: Cambridge University Press, 2003), 10.

[11] Rosaleen Duffy, "The Potential and Pitfalls of Global Environmental Governance: The Politics of Transfrontier Conservation Areas in Southern Africa," *Political Geography* 25, no. 1 (2006): 825.

[12] Manfred B. Steger, *Globalization: A Very Short Introduction*, Fifth edition, fully updated new edition, Very Short Introductions 86 (Oxford: University Press, 2020), 108.

increase in oil drilling, coal mining, as well as a plethora of other environmentally damaging policies, action to fight climate change did not halt within the US. Instead, actors at the state, county, and municipal levels coalesced in order to develop effective solutions on a local scale. Furthermore, they partnered with other global actors and organizations in order to continue to engage in agreements that would benefit planetary ecological health. These specific examples demonstrate the ways in which states no longer necessarily govern every aspect of their territorial boundaries, as actors—at least for those who sign on—follow the agreed upon rules set out in these types of global agreements. These agreements emerge from complex negotiations in an attempt to treat many of the growing global challenges. It has been accepted that without decisive and effective collective action on certain issues—such as the environment—all, or at least most, will be less well off as a result. This importance is also what is seeing an increase in environmental actors simultaneously being afforded expanding power and legitimacy. This proliferation of actors contributes to the increasingly networked public-private character of global environmental governance as this major global challenge must be treated simultaneously at the sub-state and supra-state levels while also doing so in a system that remains state-centric in a number of ways.

These examples allow us better analytical understanding of the ways global environmental governance has been and continues to evolve. For a plethora of reasons, states cannot tackle these issues alone, nor have they been able to do so effectively for some time, as seen with the widespread global failures to adequately protect our environments and fight climate change. This has led to an increase in actions ascribing authority to govern to actors previously thought unimportant as well as new types of actors altogether.[13] This authority develops in various and complex ways. It can be realized through a direct devolution of power from traditional centralized forms of government to new political units. It can also emerge from informal interactions by various actors which is in some cases legitimized through their existing actions and ongoing social relations. Duffy argues that these complex networks of actors constitute real challenges to global environmental governance, as simply understanding who exercises any degree of power or authority in any particular social relationship becomes

[13] Dan Brockington and Katherine Scholfield, "Expenditure by Conservation Nongovernmental Organizations in Sub-Saharan Africa," *Conservation Letters* 3, no. 2 (2010): 106–13.

difficult to untangle and thus operationalize.[14] The very same networks however themselves constitute global environmental governance making clear analysis and understanding of their relationships difficult to develop. Duffy suggests that "global environmental governance is closely bound up with the new networks of transnational groupings of consultants, academics and single-issue pressure groups of NGOs that constitute international epistemic communities."[15] Increasing scholarship has begun to demonstrate the way these networks and actors in conservation are expanding at a rapid pace around the world, but perhaps most of all in sub-Saharan Africa where conservation is so often intimately linked with development resources.[16]

These shifts away from govern*nment* and towards gover*nance* have meant a proliferation of community-based approaches to resource governance and conservation. Similar to community watches for local security, or community food banks for food security, community-based conservation has emerged as an approach geared towards responsibilizing residents for their local natural resources and environmental sustainability. This phenomenon realizes goals for various actors, and proponents suggest it is a way to empower communities to manage certain issues more effectively. For the state, this often consists of an outsourcing and devolution of control over resources or the management of certain environmental functions, alleviating the state of the costs associated with managing these. These costs are thus downloaded to communities, however, in many cases local residents can also benefit from these arrangements as well. Proponents argue this is an effective way to avoid tragedies of the commons and suggest collective action can be more effective when it is conducted on a small/local scale.[17] However, the process of devolution to the local level is not a simple one to one ratio of a subtraction of state power which is then added to the community in an equal and opposite empowerment. Instead, as has been discussed, public-private networks emerge in order to collaboratively enact governance and provide the types of services and efforts traditionally performed by central governments. Ryan Katz-Rosene and Matthew Paterson argue that networks of governance now include trade,

[14] Duffy, "Global Environmental Governance and the Challenge of Shadow States."

[15] Duffy, 93.

[16] Brockington and Scholfield, "Expenditure by Conservation Nongovernmental Organizations in Sub-Saharan Africa."

[17] Ernst Friedrich Schumacher, *Small Is Beautiful: Economics as If People Mattered* (New York: Harper and Row, 1979).

finance, production, interstate competition, globalization, inequalities, and the governance of all these.[18] The proliferation of networks of actors involved in governance is also seen with the evolution of CBNRM in response to climate change and the new complex partnerships that emerge as a result.

As this proliferation of actors involved in governance has occurred and both the quantity and type of actors have proliferated, new layers of environmental governance have emerged. The layering of governance that has ensued from the shifts discussed above led to ongoing political and academic debates that centre around the degree to which human activity and use of natural resources and the environment should be limited. This includes whether the state should strictly manage the conservation of environment and natural resources, or whether more integrated approaches might prove more fruitful. However, this led to debates about how conservation ought to take place, and whether restricting human use and interaction of nature in various forms of protected spaces and national parks was best or whether new approaches were needed. It is clear that by simply removing all human activity in what Brockington and Igoe refer to as 'fortress conservation' resources and natural landscapes would be protected. However, many argue that this is not only undesirable in many cases, but also doomed to fail as human populations must have a means of survival and well-being.[19] This has led to the debate as outlined by Lorraine Moore between preservationists—who advocate the preservation and non-use of resources—and utilizationists—who believe resources are best protected by being regulated through neoclassical economics and market-based approaches to conservation.[20] These two approaches also link back to the shifts in global governance as a preservationist paradigm would see central governments protect vast tracks of land and resources through sovereign decrees—for Africa this often means specifically the exclusion of local black populations. On the other hand, the utilizationist approach is intended to result in devolved control responsibilizing local residents to effectively and sustainably manage

[18] Ryan Katz-Rosene and Matthew Paterson, *Thinking Ecologically about the Global Political Economy* (London: Routledge, 2018).

[19] Doolittle, "Fortress Conservation."

[20] Lorraine Moore, "The Neoliberal Elephant: Exploring the Impacts of the Trade Ban in Ivory on the Commodification and Neoliberalisation of Elephants," *Geoforum* 42, no. 1 (2011): 51–60.

resources—which can often provide much needed livelihoods to local marginalized populations. For utilizationists:

> The notion of natural resources, including wildlife, paying their way for conservation gained popularity in the 1980s. This model of conservation, also known as 'sustainable utilisation' or 'market based approaches to conservation' is informed by the need for people to understand the value of their environment in order for them to conserve it.[21]

This was a definitive break from the past as governments attempted to move people off land—by force if necessary—in order to create national parks and similar entities that limited human use. Instead, the utilizationist approach functioned with developments in global governance that sought to increase the influence of non-state actors and increasingly to empower communities and local-level actors to give them responsibility for costs as well as benefits associated with environmental protection. While from a global perspective this shift can be seen to have been broadly occurring, this was only as a result of the agency of people on the ground. The efforts of local people and communities contributed to these shifts in highly localized ways and led to initiatives such as the Parks for People movement, transfrontier conservation areas (TFCAs), and other approaches that would see communities and resources living integrated lives.[22] These integrated approaches also led to increased efforts to govern these regional approaches including the Convention of Biodiversity, the Convention on the Protection and use of Transboundary Watercourses and International Lakes, and the Southern African Development Community (SADC) Protocol on Trade. As the former approaches to conservation led to a proliferation of actors involved in global environmental governance, the latter initiatives emerged in order to ensure that increasingly globalized resources and territories were governed effectively.

Duffy et al. demonstrate the way this proliferation of actors working towards these goals can also have more sinister outcomes.[23] These scholars, as well as Brockington and Igoe, argue that far too often marginalized people in rural sub-Saharan Africa are forced off their land or away from their

[21] Moore, 51.

[22] Maano Ramutsindela, Transfrontier Conservation in Africa: At the Confluence of Capital, Politics and Nature (CABI, 2007).

[23] Rosaleen Duffy et al., "Why We Must Question the Militarisation of Conservation," *Biological Conservation* 232, no. Complete (2019): 66–73.

traditional ways of living in the name of conservation.[24] This raises serious questions about power and authority, race and gender, and who should have say over how conservation is carried out. The political questions raised create deep tensions as the goals of many within the environmental movement globally would see protection of environments—including wildlife populations and their habitat—as necessary and worthwhile even at great costs. However, if this includes forcing marginalized populations off their land, serious issues can arise. This critique builds on the work of scholars such as Thomas Pogge that have outlined patterns that began emerging following the end of the Cold War and into the 2000s in the international and supranational governance architecture where contradictions of interests and imbalances of power often—and perhaps increasingly—pit the interests of the Global South and Global North against one another. Often these tensions and issues exacerbate underlying power imbalances which lead to a muddying of already complex phenomena. As a result, it becomes important to discuss the parallel evolution of power alongside the shifts in global environmental governance discussed above.

Power's Polymorphous Character

Whether government or governance, questions of how we order society centre around notions of power. How it is exercised, in what ways it manifests, who is able to leverage power vis-a-vis whom, and to what end? All these questions are integral to understanding the shifts that emerged within global governance in the latter part of the twentieth century. Here, I will begin by outlining some of the ways in which power has been conceptualized, in order to provide a framework for latter empirical chapters that demonstrate the ways in which CBNRM reorganizes and rearticulates power between the state and a variety of other actors including local communities, NGOs, international institutions, foreign governments, and non-state actors. Power and governance have always been complex; however, due to a number of phenomena associated with the many processes of globalization, shifts in the global political economy have increased the rate of complexification of society. As a result of this complexity, traditional notions of power and similar concepts within the social sciences have been outstripped in a number of ways. While Robert Dahl's

[24] Daniel Brockington and James Igoe, "Eviction for Conservation: A Global Overview," *Conservation and Society* 4, no. 3 (July 1, 2006): 424.

elaboration of power was frequently cited for several decades, scholars have for a number of years theorized beyond his conception.[25] According to Dahl, "A has power over B to the extent that he can get B to do something that B would not otherwise do."[26] Dahl gives the example of a professor who has power over a student to the extent that a failing grade could compel the student to study over the holidays when the student would otherwise choose not to. In a case like this, the degree of power held by the professor is measured by the difference between the probability the student would have studied versus the fear they have of failing the course. Michel Foucault is one key scholar who has since expanded on this understanding of power which is generally referred to as coercive power.[27] This type of power is that which is most obvious when we think of the state in its physical forms. This includes its passing and enforcing of laws through either force or the threat of force, as well as the inflicting of punishment for infringements of state laws.

While this conceptualization of power provides a clear outline, it is most likely to provide analytical use prior to the proliferation of the plethora of competing sets of voices and actors that influence decision making. Foucault explains what he calls absolutist and monarchical power by outlining punishment in the form of public torture in the eighteenth century. This form of power was intended to have society publicly and collectively bear witness to the absolute power of the sovereign and as a result, employ fear and coercion as a means to make subjects obey.[28] This notion of coercive power made sense to a degree when the state was seen as the preeminent actor in global politics and where power was exercised from the centre with sovereign decrees emanating towards the periphery.

Due to shifts in global governance that can be attributed to the neoliberalization of society, states' abilities have been squeezed and they have had to simultaneously do more, with less resources at their disposal. This pressure, among a confluence of other factors, has led scholars to posit that states are no longer the sole or in some cases even the principal actors

[25] Barnett and Duvall, *Power in Global Governance*; Michel Foucault, *Discipline and punish: the birth of the prison*, Social theory, 2nd edition (New York: Vintage Books, 1979); Pierre Bourdieu, "The Forms of Capital," in *Readings in Economic Sociology* (John Wiley & Sons, Ltd., 2008), 280–91.

[26] Robert A. Dahl, "The Concept of Power," *Behavioral Science* 2, no. 3 (1957): 202–3.

[27] Foucault, Discipline and punish.

[28] Foucault.

in global politics.[29] Over time, states as a result began to outsource and devolve many of the governance functions traditionally performed directly by central government apparatuses. Rita Abrahamsen argues,

> One characteristic of this mode of governance is an increasing decentralisation and pluralisation of power and decision-making centres, away from the state and towards various quasi-autonomous non-governmental organisations and institutions, so that today the 'conduct of conduct' increasingly obtains in sites 'at a distance' from the state itself.[30]

The 'conduct of conduct' as discussed here is described by Foucault as the central problem of modern government. For him, the very idea of liberal government is paradoxical as liberalism asserts the sovereignty of the free individual yet governing requires a regulation and moderation of the behaviour of individuals.[31] The conduct of conduct refers to the power to act on the actions of others. Decentralization and pluralization are intended to work within this logic and have meant increasingly that the state cannot be understood as the paramount actor in global politics. Mitchell Dean has argued that in "most social and political narratives, sovereign power is being undermined, decentred, flattened, de-territorialized, pluralized, and conceptually displaced."[32] This shift away from the state as the pre-eminent actor and referent object has a number of important impacts on understandings of International Relations. This is because traditionally most theories and concepts in the social sciences have been developed with a state-centric ontology that sees the state as the central player in the system who wields sovereign and ultimate power. While it is important to analyse these shifts, it is similarly important to acknowledge that in many cases the state still maintains and exercises a high degree of sovereign control and thus Dean and others suggest underestimating this power can present a number of risks.[33]

[29] Hall, *Emergence of Private Authority in Global Governance*, Cambridge Studies in International Relations; 85 (Cambridge, Cambridge; New York: Cambridge University Press, 2002).

[30] Rita Abrahamsen, "The Power of Partnerships in Global Governance," *Third World Quarterly* 25, no. 8 (September 1, 2004): 1459.

[31] Foucault, "The Subject and Power."

[32] Mitchell Dean, *Governing Societies: Political Perspectives on Domestic and International Rule*, Issues in Society (Maidenhead, England; Open University Press, 2007), 133.

[33] Dean, Governing Societies.

What this shift in power means is that the state-centric ontology with which many theories in the social sciences were developed ceases to keep pace with the modern complexities of society. Prior to the 1970s, this state-centric ontology developed alongside world events that were marked by catalysts such as the world wars and the Cold War—a time when this approach and the accompanying vertical stack of levels with the global on top of the national on top of the local made more analytical sense. As the era of embedded liberalism passed, however, and the period that came to be marked by neoliberal forms of governance took hold, this has ceased to be the case. Frederick Cooper outlines the fact that history is 'lumpy' and uneven.[34] This refers to the way the state and its structures have disaggregated, and some social relations have become denser while others increasingly diffuse. Cooper argues that concepts such as globalization or neoliberalism serve to smooth out this lumpiness but resultantly fail to account for a great deal.

In the contemporary period, Foucault's conception of power evolved beyond traditional understandings of sovereign power to further encapsulate disciplinary as well as biopower/governmentality[35] which were later summarized by Barnett and Duval as productive power.[36] These types of productive power are best understood in the Foucauldian sense as being everywhere, diffuse, and embodied in discourse and constantly created and enacted by the very interactions of actors.[37] For Foucault, power is not simply wielded by actors by way of episodic or sovereign acts of domination or coercion. It is instead dispersed and pervasive. It is therefore relational and uneven.[38] In relation to the current monograph, power cannot simply be seen as something that resides within the sovereign control of the state that is theirs to delegate or outsource as they see fit, but instead a much more complex set of processes and relations. Barnett and Duval posit that power works in various expressions and therefore cannot be captured in a single formulation. They suggest there are four types of

[34] Frederick Cooper, "What Is the Concept of Globalization Good for? An African Historians Perspective," *African Affairs* 100, no. 399 (2001): 189–213.

[35] Michel Foucault, "Power/Knowledge: Selected Interviews and Other Writings," Penguin Random House Canada, 1977.

[36] Barnett and Duvall, Power in Global Governance.

[37] Michel Foucault, Power/Knowledge: Selected Interviews and Other Writings, 1972-1977, 1st ed. (New York: Vintage, 1980).

[38] Michel Foucault, *"Society Must Be Defended": Lectures at the Collège de France, 1975-76*, 1st Picador pbk. ed., (New York: Picador, 2003), 43.

power: compulsory, institutional, structural, and productive.[39] These understandings of multiple expressions of power and the way it can be produced and leveraged to achieve ends are contrary to so much work that has been done within the academe to try and neatly conceptualize and define power. This kind of oversimplification of the complex concept of power is particularly seen with Robert Dahl's early definition as well as realists such as E.H. Carr and Brodie and Dunn.[40] This book analyses the ways in which power relations between various actors contribute to certain outcomes over others. These expressions of power can be produced by various actors in complex ways and are greatly influenced by developments in global environmental governance.

By relying only on traditional definitions of power, based on understandings of states as coherent and paramount structures within the international, we fail to account for a great deal. Instead Barnett and Duvall argue a "healthy recognition of power's polymorphous character, and a willingness to look for connections between these different forms enhances and deepens our understanding of international politics."[41] Cooper calls for 'lumpier' understandings of difficult concepts such as globalization that attempt to smooth over a great deal of complexity, while Larner and Walters similarly claim that we must accept 'messier' analyses that simultaneously provide more in the way of accuracy.[42]

As these moves away from notions of sovereign power have proliferated and become increasingly accepted, there has been a shift towards conceptions of productive power. Along with these developments not only do the key actors shift—away from the state—but also the forms of power evolve away from instrumentalism and towards more relational and productive forms of power. These are two key changes that this book analyses. Coercive power has necessary limits and demands a great deal of resources as it is intimate and must involve a certain degree of physical connection between ruler and subject. Unlike sovereign or coercive power, productive power is not

[39] Barnett and Duvall, Power in Global Governance.

[40] For example see: E. H. Carr, The Twenty Years' Crisis, 1919-1939: Reissued with a New Preface from Michael Cox (Springer, 2016); Bernard Brodie and Frederick Sherwood Dunn, The Absolute Weapon: Atomic Power and World Order (Harcourt, Brace and Company, 1946).

[41] Barnett and Duvall, *Power in Global Governance*, 4.

[42] Wendy Larner and William Walters, "Globalization as Governmentality," *Alternatives: Global, Local, Political* 29, no. 5 (2004): 495–514.

> the brute fact of the domination of the one over the many, or of one group over another, but the multiple forms of domination that can be exercised in society; so not the king in his central position, but subjects in their reciprocal relations; not sovereignty in its one edifice, but the multiple subjugations that take place and function within the social body.[43]

Exercising sovereign power effectively becomes impractical, if not impossible as societies grow, and thus, Foucault discusses the shift to disciplinary power through his discussion of Jeremy Bentham's Panopticon. The Panopticon is described as a central tower staffed by prison guards who watch over all prisoners whose cells face inward towards the guard tower knowing their every move *can* be seen, yet never knowing for sure if at any moment they *are* being watched. Foucault describes this as how power is increasingly exercised, as every action of social beings within a society cannot be effectively policed. Instead, the sovereign passes laws and creates a dense web of rules and regulations by which people must operate within their everyday life as well as almost every action.

This power becomes productive as people fear the threat that their ability to operate relatively freely within this societal set of rules can be taken away and their physical bodies subject to a removal of any number of freedoms through the penal system—generally referring to prison in its different forms.[44] Foucault's conception of power evolved over time and his earlier work we only see sovereign and disciplinary power while later he introduces a third category of power which is even subtler. This form of power which is variously referred to as biopower, governmentality, and/or security is that which sees power exercised in perhaps its most capillary and relational way and is that which is furthest from direct sovereign actions.[45] This form of power is also productive like disciplinary power but is more subtle, pervasive, and operationalized based on the social relations of subjects within society, rather than through hierarchical relations.[46] For Foucault, governmentality refers to the conduct of conduct, which in its simplest form entails the control of how people conduct themselves. This is something that he explored from smaller institutions right up to the governance of large modern societies and the various and limitless power

[43] Foucault, Society Must Be Defended, 27.

[44] Foucault, Discipline and punish.

[45] Michel Foucault, *Security, Territory, Population: Lectures at the Collège de France, 1977-78* (Houndmills, Basingstoke, Hampshire; Palgrave Macmillan, 2007).

[46] Foucault, Power/Knowledge.

relations that exist within them and enable them to function. He looked deep into the way governments function, the activities they pursue, and how their actions are experienced by those in society. His analysis of this form of power includes questions around who can govern, who can be governed, and what governing is or can consist of.[47] It is this form of power which is built on by a number of scholars later on that is most prevalent in many studies of global governance and a key form of power interrogated in this book.

While Foucault generally writes on sovereign, disciplinary, and biopower/governmentality, both disciplinary and biopower/governmentality are productive forms of power which are usefully conceptualized later by Barnett and Duvall as productive power.[48] It is these forms of productive power that are subtle and often indirect which this book will focus on predominantly in the empirical chapters.[49] Importantly, productive power is about much more than preventing subjects from committing crime and is intended to be useful, productive, and effective in responsibilizing citizens for their own well-being. Going back to our discussion of Dahl, the example of a professor getting a student to study proves a useful example. Professors are—at least rarely—seen as tyrannical authoritarians who exercise their will through coercive force and domination. Instead, the relationship between teacher and student is more often seen as one where the teacher has a legitimate claim to power based on their knowledge and experience gained through education, university degrees, and their position as professor. Students generally accept this power as legitimate, and as a result are incentivized to complete their work, or study over a holiday as a result of the potential for punishment or award (in the form of good or bad grades) the teacher will offer in response to the results achieved through the student's work. Evolutions in global governance have resulted in a similar process with the ways in which power is produced more broadly in society between the state and its subjects. This shift away from sovereign or coercive forms of power leads scholars to argue that the state is no longer the sole, or in many cases even the principal, source of authority in either the domestic or the international arenas.[50]

[47] Foucault.

[48] Barnett and Duvall, Power in Global Governance.

[49] Barnett and Duvall.

[50] Wendy Larner, "Neo-Liberalism: Policy, Ideology, Governmentality," *Studies in Political Economy*, no. 63 (2000): 5–25; Rosenau, "Governance in the Twenty-First Century," January 1, 1995; William Walters, "Some Critical Notes on 'Governance,'" *Studies in Political Economy* 73 (April 1, 2004): 27–46.

While sovereign power is exercised in a direct top-down fashion, the type of productive power that Foucault refers to as governmentality or biopower is operationalized between individuals within society as they form various relationships across multiple contexts.[51] This occurs in virtually every facet of society as the knowledge or discourses of medicine, education, and punishment permeate hospitals, schools, and prisons as a way to discipline and structure those within them.[52] Importantly for Foucault, while society has and continues to evolve alongside these forms of power, as productive power has become more pervasive, it has not simply displaced other forms of sovereign power. Instead, productive power "infiltrated the others, sometimes undermining them, but serving as an intermediary between them, linking them together, extending them and above all making it possible to bring the effects of power to the most minute and distant elements."[53] While they have always done so to some degree, especially since the end of the Cold War, states increasingly rely on and work to develop forms of productive power which permeate society in order to realize their interests with the use of limited resources.[54] The next section will detail what these various forms of community conservation can look like while placing a specific focus on CBNRM as one method of doing so.

Community-Based Natural Resource Management

CBNRM as a product of the shifts in global environmental governance must be understood in the context of Africa's history of poor resource governance and the challenges of sustainable development. It traces its origins within development thinking in the 1980s when dominant discourse came to focus on participation, ownership, and empowerment. These shifts in both discourse and approaches to conservation were intended to prevent large-scale corruption and ensure inclusive socio-economic development for local populations.[55] Beginning in the 1980s, the shift from a preservationist conservation paradigm to a more integrated approach began to recognize the importance and necessity for the promotion and empowerment of communities. This integrated approach was

[51] Foucault, Discipline and punish, 146.

[52] Foucault, Society Must Be Defended, 241.

[53] Foucault, Discipline and punish, 216.

[54] Barnett and Duvall, Power in Global Governance.

[55] Fabricius and Collins, "Community-Based Natural Resource Management."

also developed in contrast to the push by many to replace customary land tenure with a Western type of legal registration process. Ambreena Manji argues that those pushing for this Western approach sought to further neoliberalize countries already hurt by neoliberal policies and that pursuing this path would lead to further issues.[56] Integrated and community-based approaches to land reform then sought to reform land by keeping certain customary land tenure techniques and pairing communal forms of land and resource management. The paradigm shift suggested that this empowerment could be accomplished by linking economic and social development to natural resource management and local environmental entitlements.[57] Together, scholars and practitioners have come to develop the groundwork for participatory and community-based approaches that have proliferated in recent years. For many, "community-based management has become the rallying call of green politics, indeed is held as a prerequisite for environmental sustainability."[58] This ethos also coincides with a general trend in rural conservation and development studies that suggests communities should be included in decisions that affect them, especially as it is these communities who best understand local nuances. It is from these developments that emerged community-based conservation approaches as well as initiatives such as the Parks and People Programme and TFCAs as outlined above.[59] While these various approaches to conservation emerged within the same paradigm and overlap in many ways, this book will focus squarely on CBNRM as one specific example that emerges from these many approaches.

Academic discussions and theories of development must be understood as intimately intertwined with the practices of global environmental governance. One necessarily impacts the other and shifts in one create reactions and outcomes in the next. Academia and the social and political world do not exist in silos similarly to how the private and public sectors cannot be understood as heterologous. Thus, as the complexification of

[56] Ambreena S. Manji, The Politics of Land Reform in Africa: From Communal Tenure to Free Markets (Zed Books, 2006).

[57] Melissa Leach, Robin Mearns, and Ian Scoones, "Environmental Entitlements: Dynamics and Institutions in Community-Based Natural Resource Management," *World Development* 27, no. 2 (1999): 225–47.

[58] Michael J. Watts, "Antinomies of Community: Some Thoughts on Geography, Resources and Empire," *Transactions of the Institute of British Geographers* 29, no. 2 (2004): 196.

[59] M. Ramutsindela, Parks and People in Postcolonial Societies: Experiences in Southern Africa (Springer Science & Business Media, 2004).

society led to the proliferation of actors and the outsourcing of state functions, development discourse was similarly evolving and came to theorize these shifts; in turn reifying them and contributing to a move away from state-centricity. This is key to understanding the emergence of CBNRM within these shifts in global politics and development discourse.

While CBNRM has a long history, it emerged in a systematic fashion and grew in popularity as dominant thinking in international development came to focus on community empowerment as the key principles in development discourse shifted focus towards participation and ownership.[60] Swatuk argued at the time that "tantamount to a new 'truth' of development, 'participation' is a political cross-conditionality in current North-South inter-state relations."[61] After decades of failed natural resource governance, these ideas entered the mainstream, suggesting that effective conservation solutions demanded widespread participation of local people and communities throughout the entire planning, implementation, and ongoing management phase of strategies.[62] Madhusree Sekhar summarized this by arguing that "participatory natural resources management has become the new catchphrase in tackling environmental problems, particularly those arising from the use of common property resources."[63]

Not only was the practice of participation seen as necessary, but eventually also ownership so that communities could be empowered and they would importantly feel a vested interest in their success.[64] CBNRM emerged with these two principles of participation and ownership at its core.[65] These two concepts aimed to place those that development initiatives were intended to help at the centre of efforts at solutions. The idea was that this would eliminate the type of cookie-cutter approaches to development that had contributed to decades of failure, and which saw the international community decide what would be best to achieve

[60] Aram Ziai, Development Discourse and Global History: From Colonialism to the Sustainable Development Goals (Routledge, 2015).

[61] Swatuk, "From 'Project' to 'Context,'" 99.

[62] Claire Mercer, "Performing Partnership: Civil Society and the Illusions of Good Governance in Tanzania," *Political Geography* 22, no. 7 (2003): 741–63.

[63] Bhaskar Vira and Roger Jeffery, "Analytical Issues in Participatory Natural Resource Management," Global Issues Series (Houndmills, Basingstoke, Hampshire; Palgrave, 2001), 87.

[64] Joseph E. Stiglitz, Senior Vice President and Chief Economist, 1998.

[65] Masumi Owa, "Revisiting the Paris Declaration Agenda—an Inclusive, Realistic Orientation for Aid Effectiveness," *Development in Practice* 21, no. 7 (2011): 987–98.

development—which was also defined in a particular way—and then implement similar programmes in diverse locations. Participation aimed to shift this paradigm in order to insert development recipients into the process of planning and implementation itself. This paradigm further evolved to include ownership, so that local populations could take on a central role in projects and eventually have full managerial decision-making capacity so that they were implicated in their own developmental success or failure.[66] CBNRM is intended to achieve these concepts by empowering local residents, to give them a stake in the protection of their natural resources. However, Andrea Cornwall and Karen Brock argue,

> In the fast-moving world of development policy, buzzwords play an important part in framing solutions. Today's development orthodoxies are captured in a seductive mix of such words… [Cornwall and Brock] show how words that once spoke of politics and power have come to be reconfigured in the service of today's one-size-fits-all development recipes, spun into an apoliticised form that everyone can agree with.[67]

As a result, it becomes important to unpack both what these concepts are employed to achieve and what they are intended to gloss over.

Community-based forms of conservation stem from the increasingly popular notion that conservation and economic development can, and for some must, be achieved simultaneously. This shift contrasts with traditional forms of development and resource management, which have often been in direct conflict with environmental conservation and sustainability. Community conservation seeks to reformulate these traditionally antagonistic relationships by placing an emphasis on the role of local residents in decision making regarding natural resources. These forms of resource governance emphasize conservation by, for, and with the local community.[68]

CBNRM has become one of the most prolific and established forms of community conservation. It intends to transfer

[66] Aram Ziai, Development Discourse and Global History: From Colonialism to the Sustainable Development Goals (Routledge, 2015).

[67] Andrea Cornwall and Karen Brock, "What Do Buzzwords Do for Development Policy? A Critical Look at 'Participation', 'Empowerment' and 'Poverty Reduction,'" *Third World Quarterly* 26, no. 7 (October 1, 2005): 1043.

[68] Hoole, "Place-Power-Prognosis."

administrative rights over wildlife, forests and water sources to agents at the local level with three aims: (a) to ensure sustainable management of natural resources in rural areas, (b) to democratize decision-making in rural communities, and (c) to ensure that the benefits of natural resources can be harvested at the local level.[69]

Its intended ethos is that humans and nature can coexist peacefully, and prosperity can be a result of a healthy relationship between the two.[70] This ethos varies from other forms of conservation that seek to limit human impact on the natural environment, as it suggests people must be empowered to achieve economic self-sufficiency in a way of their own choosing. In developing regions in general, and in Africa in particular, this has been an exigent point as expanding population pressures mixed with the progressing detrimental effects of climate change have increasingly strained land and local environments. With over 60 per cent of the African continent engaging in some form of subsistence agriculture as a means of survival, any approach that seeks to limit the ability of people to prosper from the land is certain to provoke resentment and discontent.[71]

CBNRM and similar forms of community-based conservation are not only intended to engender economically sustainable environmental conservation, but also touted by many most specifically for their community empowerment through local institutional development.[72] Several studies suggest that "since development and poverty reduction are multidimensional, perhaps the greatest value of CBNRM is political transformation through democratization, equity, transparency and accountability."[73] Commons scholarship has long been developed by leading scholars such as Elinor Ostrom that seeks to deal with common-pool resources such as

[69] Michael Bollig and Diego Menestrey Schwieger, "Fragmentation, Cooperation and Power: Institutional Dynamics in Natural Resource Governance in North-Western Namibia," *Human Ecology* 42, no. 2 (2014): 167.

[70] Jonathan Otto et al., *Natural Connections: Perspectives In Community-Based Conservation*, ed. David Western and Michael Wright, 2 edition (Washington, D.C.: Island Press, 1994).

[71] Makhtar Diop, "Foresight Africa 2016: Banking on Agriculture for Africa's Future," *Brookings* (blog), 2016.

[72] Adam Cruise and Sasada Izzy, "Investigation into the Efficacy of Namibia's Wildlife Conservation Model as It Relates to African Elephants (Loxodonta Africana)," 2021.

[73] Brian Child and Grenville Barnes, "The Conceptual Evolution and Practice of Community-Based Natural Resource Management in Southern Africa: Past, Present and Future," *Environmental Conservation* 37, no. 3 (2010): 284.

lakes, oceans, irrigation systems, fishing grounds, forests, the internet, the stratosphere, and wildlife populations.[74] Commons dilemmas arise when one person's consumption of a given resource limits the ability of others to consume the same resource. Leading commons scholarship has stressed that local levels of collective action—such as with the responsible management of local resources—are necessarily linked to more effective levels of social and political organization in the highly globalized world.[75]

As Michael Bollig suggests, rather than attempting to remedy the political issues discussed above through privatization or extended state control, it was through determined efforts to reorganize resource governance that various forms of community-based conservation were developed.[76] Brian Child and Grenville Barnes previously argued that CBNRM served to marketize valuable commons resources "so that CBNRM's comparative economic advantage is reflected in landholder and community incentives."[77] Bollig further unpacks CBNRM by positing:

> The argument was simple: once natural resources were valorized people would take care to manage them rationally to ensure continuous returns; valorization, commoditization and an opening up to emergent markets would ensure increased local incomes and a diversification of livelihoods. Two measures were necessary to achieve this: local institutions needed to be efficiently organized in order to ensure governance and surveillance of natural resources and a gearing between private sector investors and communities needed to be organized to facilitate marketization and income generation.[78]

To do so communities would have to manage resources sustainably in order to ensure continuous returns rather than engaging in short-term consumptive uses that provide far less in net benefits. By valourizing resources, proponents argue, commoditization would occur, leading to an

[74] Ostrom, Governing the Commons.

[75] Fikret Berkes, "Commons in a Multi-Level World," *International Journal of the Commons* 2, no. 1 (November 1, 2007): 1–6; D. Cash et al., "Scale and Cross-Scale Dynamics: Governance and Information in a Multilevel World," *Ecology and Society* 11, no. 2 (2006).

[76] Michael Bollig, "Towards an Arid Eden? Boundary Making, Governance and Benefit Sharing and the Political Ecology of the 'New Commons' of Kunene Region, Northern Namibia," *International Journal of the Commons* 10, no. 2 (2016): 771–99.

[77] Child and Barnes, "The Conceptual Evolution and Practice of Community-Based Natural Resource Management in Southern Africa," 283.

[78] Bollig, "Towards an Arid Eden?," 772.

opening up of multiple market areas that could lead to positive economic offshoots and a continued impetus for long-term sustainable management.[79] Louise Fortmann, Emery Roe, and Michel van Eaten suggest CBNRM is based on four assumptions: (1) local people are reasonably knowledgeable about local ecosystems, indeed often more knowledgeable than most outside 'experts' are likely to be; (2) a benefits flow can be created from management activities that significantly outweighs the costs of coexisting with the resource or the benefits foregone from other uses or management strategies; (3) a group capable of implementing management strategies exists; and (4) control over the resources will be devolved to the community.[80]

CBNRM involves the collective use and management of natural resources in rural areas by a group of individuals who self-identify as a defined and distinct social entity with a specific identity using communally owned and operated facilities. Programmes are intended to be flexible and designed for voluntary participation by self-identified communities.[81] Benefits can take many forms but generally include direct financial gain from the sale of resources, ecotourism, manifold employment opportunities, and the development of related local entrepreneurial activities. Programmes are generally enabled through the provision of technical and financial resources and assistance to communities mainly by an assemblage of NGOs that seek to participate in order to maximize opportunities for success. While CBNRM can take many forms, ecotourism has become the most prolific model across Southern Africa. It is suggested that "ecotourism is promoted by governments of the North and South, international

[79] Arun Agrawal and Clark C. Gibson, "Enchantment and Disenchantment: The Role of Community in Natural Resource Conservation," *World Development* 27, no. 4 (1999): 629–49; Fortmann, Roe, and van Eeten, "At the Threshold between Governance and Management"; Leach, Mearns, and Scoones, "Environmental Entitlements"; Diego A. Menestrey Schwieger, "An Ethnographic Analysis of the Role of Power in Institutional Arrangements: Borehole Cost Recovery within a Pastoral Community in North-Western Namibia," *Environmental Policy and Governance* 25, no. 4 (2015): 258–69; Ostrom, *Governing the Commons*; Arne Rückert, "Producing Neoliberal Hegemony? A Neo-Gramscian Analysis of the Poverty Reduction Strategy Paper (PRSP) in Nicaragua," *Studies in Political Economy* 79, no. 79 (2007); Michael Schnegg, "Lost in Translation: State Policies and Micro-Politics of Water Governance in Namibia," *Human Ecology* 44, no. 2 (April 1, 2016): 245–55.

[80] Fortmann, Roe, and van Eeten, "At the Threshold between Governance and Management," 171.

[81] Boudreaux, "A New Call of the Wild."

lending institutions, and private business, as an ideal development strategy that combines economic growth with environmental conservation."[82] While CBNRM was developed to deal with many commons resource use issues, it has predominantly taken the form of wildlife management through ecotourism ventures. CBNRM in general, and wildlife conservation specifically, makes sense in the Southern African context as

> wildlife, and particularly the more charismatic megafauna of Africa, is Africa's global comparative competitive advantage over the rest of the planet. While virtually every country on Earth has cattle, sheep, and goats, only the continent of Africa has the variety and spectacle of wildlife that makes it stand out on the global landscape.[83]

This is to say that wildlife in Africa is more valuable than in many other places. Furthermore, land in many other settings is more agriculturally productive, as well as being better situated to have access to lucrative markets, making agriculture generally more productive than wildlife management in many other regions with the inverse being true in the arid climates of Southern Africa.

Ecotourism in this context generally involves the drafting of national legislation that lays the framework for the responsible devolution of resources from national governments to local communities who must effectively organize to develop local-level institutional capacity, and resource-sharing strategies. From there, communities must develop defined conservancies in which they are intended to have decision-making capacity and control over wildlife and their habitat while placing a focus on local employment and entrepreneurial activities.[84] This can take the form of

> lodges [that] provide hundreds of full and part-time jobs in areas where there are few employment options other than subsistence farming. Joint venture agreements require lodge operators to employ and train local people whenever possible… Each helps some local people develop skills and abilities that expand their possibilities of employment and their abilities to

[82] Duffy, "Global Environmental Governance and the Challenge of Shadow States," 96.

[83] Chris Brown, "Opinion: Ecologist Responds to Guardian Newspaper Article against Trophy Hunting," *Africa Geographic* (blog), January 2, 2019, 2.

[84] Michael Schnegg, "Institutional Multiplexity: Social Networks and Community-Based Natural Resource Management," *Sustainability Science* 13, no. 4 (July 1, 2018): 1017–30.

run businesses and manage wildlife as business opportunities expand, the local economy improves.[85]

Joint-venture operations are set up with local and international NGOs as well as various private sector enterprises to agree on leases to operate in the conservancy and draw in international tourists. The ecotourism industry aims to attract international 'up-market' clientele. Revenues from tourists are intended to fund conservancy operations as well as the many offshoots international tourism creates, such as within the food and service industry, local artisanal shops and wares as well as increased traffic for local enterprises.[86] As tourists are attracted based on the stunning wildlife, habitat, and scenery it then becomes a major focus of conservancies to effectively organize to protect and promote these natural endowments.

While attractive for conservationists, there are also potential limitations associated with these foundational principles. Some scholars have denounced these ideals as simply the newest buzzwords in development discourse which seek to be so all-encompassing as to mean very little at all.[87] It has also been similarly argued that the concept of participation serves only to gloss over the latest version of neocolonialism.[88] This is a specific issue that Thembela Kepe and Ruth Hall discuss in South Africa as they discuss the question of land distribution and whether certain practices are actually leading to recolonization that limits land for certain groups which has become an increasing concern across Southern Africa and perhaps specifically in South Africa.[89] After decades of failed development interventions in which the Global North seemed to be continuing its domination and oppressive tactics this shift to community empowerment has been seen sceptically by some. While the ethos of CBNRM intends for the empowerment of communities, there is also the ongoing overarching presence of both domestic and foreign state powers as well as a number of international actors who oversee the projects and activities. This has led to continued caution for many as there have been fears this could become the

[85] Boudreaux, "A New Call of the Wild," 315.

[86] Child, "The Practice and Principles of Community-Based Wildlife Management in Zimbabwe."

[87] James Fairhead, Melissa Leach, and Ian Scoones, *Green Grabbing: A New Appropriation of Nature*, Critical Agrarian Studies (London: Routledge, 2013).

[88] Bill Cooke and Uma Kothari, *Participation: The New Tyranny?* (Zed Books, 2001).

[89] Thembela Kepe and Ruth Hall, "Land Redistribution in South Africa: Towards Decolonisation or Recolonisation?," *Politikon* 45, no. 1 (January 2, 2018): 128–37.

most recent iteration of neocolonial policies that results in little change in practice, despite the fancy rhetoric.

Part of these fears emerges as participation can prove to be a challenging piece of the puzzle to effectively integrate. Abrahamsen argues that efforts at participation often lead to what can be viewed as consultation, as actors arrive with ideas that become difficult not to instil in development efforts on the ground.[90] Consultation might see people informed about pre-determined plans, and perhaps offer some input in the development phase of a project, but fundamentally be removed from influencing the course of events or strategies chosen as will be seen below with the case study of Namibia. This is indeed where politics itself resides, within deliberation, difference of opinions, and struggles for power and resources. Put another way then, politics cannot be removed from a situation and politics and power struggles influence so many pieces of every conservation puzzle.

Gaps and Disputes

This chapter has situated CBNRM within transformations in global governance and demonstrated how it is an articulation of more decentred, public-private, global-local forms of rule and policymaking. It has also unpacked the evolving debates about CBNRM and has revealed the various causes and consequences scholars have pointed to for diminished success in recent years. Moreover, the chapter has unpacked questions of politics and the local-global relations involved in CBNRM. A complex picture emerges that demonstrates that success in some senses has been waning; however, it seems that to some degree success has also been judged differently over time. While there is evidence that suggests success in attaining the espoused goals of CBNRM has been ebbing, it becomes clear that it is specifically many within rural communities who have failed to gain the benefits they felt they were promised and that gender and race issues persist.

Since its inception in the 1990s, CBNRM has contributed to a number of sustainable development goals for Namibia. However, it is clear that the programmes have fallen short of their intended goals in a number of ways. These findings are important as they outline challenges facing the CBNRM programme but analysed together, a clear omission is discussion of the

[90] Abrahamsen, "The Power of Partnerships in Global Governance"; Mercer, "Performing Partnership"; Stiglitz, *Senior Vice President and Chief Economist*.

detrimental effects climate change is having on the programmes. This book seeks to fill this gap by having the impacts of climate change as central to the analysis in order to understand the ways in which it influences and effects the governance of CBNRM while arguing that further policies must be crafted to acknowledge these changing realities.

Beyond climate change as an important gap in the literature, scholars widely agree that CBNRM works best where devolution occurs in more than just name and where small communities are placed at the heart of resource governance.[91] There is consensus in the literature that community empowerment is key for success, but it differentiates on the degree to which this important goal is attained.[92] Several studies point to the way the state maintains control and, despite outsourcing some power to communities, the government maintains final say over the direction of developments in the conservancies.[93] Many argue that communities often remain unheard and unable to effectively exercise agency however, and it is unclear to what degree this is an accurate portrayal of the politics involved.[94]

Scholars have further pointed to the many partnerships, resources, and the training necessary to launch these projects and to continue engaging in efforts at attaining success.[95] Ndeinoma, Wiersum, and Arts refer to this as the public-private governance network,[96] while most other studies relating to CBNRM discuss the influence the network of NGOs and other international actors maintain on the projects from implementation

[91] Hulme and Murphree, *African Wildlife and Livelihoods*; Brian Jones and L. Chris Weaver, "CBNRM in Namibia: Growth, Trends, Lessons and Constraints," in *Evolution and Innovation in Wildlife Conservation* (Routledge, 2008).

[92] Boudreaux, "A New Call of the Wild"; Lelani M. Mannetti et al., "Expanding the Protected Area Network in Namibia An Institutional Analysis," *Ecosystem Services* 28, no. Part B (2017): 207–18; Michael Schnegg and Richard Dimba Kiaka, "Subsidized Elephants: Community-Based Resource Governance and Environmental (in)Justice in Namibia," *Geoforum* 93, no. Complete (2018): 105–15.

[93] Boudreaux, "A New Call of the Wild"; Jen Dyer et al., "Assessing Participatory Practices in Community-Based Natural Resource Management: Experiences in Community Engagement from Southern Africa," *Journal of Environmental Management* 137, no. Complete (2014): 137–45.

[94] Schnegg and Kiaka, "Subsidized Elephants."

[95] Kathleen A. Galvin et al., "African Community-Based Conservancies: Innovative Governance for Whom?," in *Nomad-State Relationships in International Relations*, ed. Jamie Levin (Cham: Springer International Publishing, 2020), 147–72.

[96] Ndeinoma, Wiersum, and Arts, "The Governance of Indigenous Natural Products in Namibia."

onward.[97] Interestingly, it is only mentioned in some of the most recent work by Schnegg and Schnegg and Kiaka that NGOs and other international partners had a substantial role in the drafting of the initial legislation that provided the legal framework for CBNRM in Namibia.[98] Ongoing NGO participation and partnership in CBNRM remain understudied and are imperative to understanding and analysing how governance is impacted. While some scholars point to these power dynamics and some of the intricacies of these relationships,[99] there has yet to be analysis developed that provides adequate understanding of the politics involved from the local to the global levels. As NGOs play a central role in CBNRM projects it becomes essential to understand where they procure their funding and how they produce power and legitimacy as well as what degree of control these funders maintain on NGO activities. These webs of connectivity that link the local to the global have important—if not always obvious—effects on the way conservation can play out on the ground. Existing studies mention some of the roles NGOs and other international actors play,[100] but further analysis into the social relationships, power asymmetries, and political motives of these actors is required for unpacking the complexities of this form of resource governance. More analysis is needed to determine where these various NGOs and outside actors emerge from and the further layer of political influence that exists. Beyond this, it becomes important to interrogate questions of race and gender and who sits in various seats of power and thus dictate outcomes.

There has yet to be adequate attention in the literature regarding the politics of CBNRM and the way it is given rise to by transformations in global environmental governance. These transformations as described in this chapter restructure power, and CBNRM is an expression of these evolutions in governance. These gaps are partly a result of the nature of existing approaches as many of the early studies on CBNRM tended to approach it as an apolitical policy shift. However, as has begun to be demonstrated by some of the more recent literature on CBNRM, there is always a strong global dynamic to these projects. I build on these to develop my analysis in this book. To analyse this form of

[97] Schnegg and Kiaka, "Subsidized Elephants."

[98] Schnegg, "Lost in Translation"; Schnegg and Kiaka, "Subsidized Elephants."

[99] Michael Schnegg, Michael Bollig, and Theresa Linke, "Moral Equality and Success of Common-Pool Water Governance in Namibia," *Ambio* 45, no. 5 (2016): 581–90.

[100] Duffy, "The Potential and Pitfalls of Global Environmental Governance."

community-based conservation that I argue cannot be adequately understood outside global power relations, the tools of International Relations can capture new dimensions that account for the increasing interconnectedness of politics. CBNRM must be adequately analysed beyond a policy prescription and its enabling mechanisms traced in order to begin to comprehend and conceptualize the extent of it in relation to broader questions of global environmental governance. Only in doing so can we begin to uncover the ways in which CBNRM has evolved in response to climate change and account for which actors are driving the change and decision making. The next chapter will outline how a global assemblage approach can be applied to further our understanding of CBNRM as an expression of global environmental governance.

CHAPTER 3

Assembling Evolutions in Governance

In this chapter, I outline my theoretical framework, drawing on recent International Relations literature on global assemblages. I provide a brief overview of the more philosophical foundations of assemblage thinking, prior to discussing some of the ways it has evolved, specifically within the discipline of IR as well as ways it has been employed to understand various complex policy formations in the social world. The chapter goes on to discuss the ways scholars have developed global assemblages beyond a traditional theoretical framework and ontology, to also serve as a methodological approach for unpacking and analysing complex phenomenon. This methodology section will include an elaboration of Namibia as my case study, my approach to fieldwork, and process of data analysis and triangulation. Namibia was chosen for this study as the crucial case design as it is recognized as a global leader in CBNRM due to having devoted the highest percentage of land to this form of resource governance on earth, as well as having a recently drafted and progressive constitution with stipulations for environmental conservation written throughout its sections.

> *Global processes also take place at subnational levels, hereby disrupting the notion of mutually exclusive domains for the national and global.*[1]

[1] Saskia Sassen, "When National Territory Is Home to the Global: Old Borders to Novel Borderings," *New Political Economy* 10, no. 4 (2005): 523.

A. Heffernan, *The Global Politics of Local Conservation*, Environmental Politics and Theory,
https://doi.org/10.1007/978-3-031-24177-2_3

Evolutions in global governance have created a need to abandon dichotomies of old, yet in doing so we must be left with tools for useful and rigorous analysis. CBNRM is one expression of the re-ordering of power between various actors as well as what were previously understood as the local, national, and global and public and private spheres. The results are a number of complex multi-scalar connections and understanding these emerging phenomena requires analytical tools and methodological approaches that outstrip many traditional theories within the social sciences. Evolutions in global governance have also meant that power has become even more dispersed and pervasive than before—though it has likely always been more diffuse than traditionally understood. Rather than natural resources managed directly by the state, CBNRM provides a vehicle for power to be devolved to communities that interact with a number of actors resulting in complex and layered governance arrangements. As power is devolved from central governments, a variety of both sub-state and non-state actors take on new roles and, in many cases, begin to perform state-like practices. The result is that decision making and authority are constantly shifting based on the social interactions of competing and often heterogeneous actors. Understanding the ways in which their complex interactions have resulted in evolutions to the programme becomes important as it is clear they do not act within silos. Developing tools and approaches to analyse these evolving power relations is key and this chapter will lay out my methodological approach for doing so.

Thomas Kwasi Tieku and Tim Murithi argue that "a good theoretical account of Africa's international relations must at the very least recognize that Africa's international relations is distinct from international politics of the so-called great powers, which has been the main focus of traditional International Relations."[2] Assemblage thinking is one approach that enables scholars to analyse changes in global governance over time by unpacking foundational concepts of the discipline of International Relations such as state, sovereignty, and borders, as well as to move beyond the traditional levels of analysis approach. Moving beyond the predominant theories in the social sciences which have been developed with state—and often Western—centric ontologies enables us to engage with complex phenomena with a flattened ontology. This approach serves to prevent much of the reification that traditional theories can lead to and allows us

[2] Thomas Kwasi Tieku, "Theoretical Approaches to Africa's International Relations," in *Handbook of Africa's International Relations*, 1st ed. (Routledge, 2014), 3.

to enter the field without pre-determined notions of what we expect or seek to find. Building on global assemblage thinking from scholars of International Relations, global policy assemblage literature provides a particularly salient approach for analysing CBNRM and the way it adapts to changing realities over time.

Assemblage Thinking as a Means of Analysis

Global assemblages as an approach have been employed to understand the complexities of a range of modern phenomena. These include more abstract or big picture issues such as the formation of cities,[3] international relations,[4] global security assemblages,[5] and various other global assemblages.[6] It also includes policy formations such as the practices of pig farming in the European Union,[7] land-use planning in Tanzania,[8] Indonesian forest management,[9] policies on creative industries in New Zealand,[10] and even the Paralympics.[11] Jane Bennett has also contributed an operationalized account of an assemblage approach in her discussion of the 2003 North American blackout, providing an empirical analysis of the way in which complex arrangements of both human and non-human entities

[3] Eugene McCann and Kevin Ward, *Mobile Urbanism: Cities and Policymaking in the Global Age* (U of Minnesota Press, 2011).

[4] Acuto and Curtis, "Assemblage Thinking and International Relations."

[5] Rita Abrahamsen and Michael C. Williams, "Security Privatization and Global Security Assemblages," *The Brown Journal of World Affairs* 18, no. 1 (2011): 171–80.

[6] Stephen J. Collier, "Assemblages and the Conduct of Inquiry," in *Reassembling International Theory: Assemblage Thinking and International Relations*, ed. Michele Acuto and Simon Curtis (London: Palgrave Macmillan UK, 2014), 32–38.

[7] Elizabeth C. Dunn, "Standards and Person-Making in East Central Europe," in *Global Assemblages* (John Wiley & Sons, Ltd., 2008), 173–93.

[8] Chris Huggins, "Land-Use Planning, Digital Technologies, and Environmental Conservation in Tanzania," *The Journal of Environment & Development* 27, no. 2 (June 1, 2018): 210–35.

[9] Tania Murray Li, "Practices of Assemblage and Community Forest Management," *Economy and Society* 36, no. 2 (May 1, 2007): 263–93.

[10] Russell Prince, "Policy Transfer as Policy Assemblage: Making Policy for the Creative Industries in New Zealand," *Environment and Planning. A*, Environment and Planning A, 42, no. 1 (2010): 169–86.

[11] Andrew Heffernan, "Paralympic Assemblages Globalizing International Relations: An Autoethnographic Account of Global Politics at the Paralympic Games," *Sport in Society* 0, no. 0 (June 5, 2020): 1–14.

assemble to create functioning—or in the case she presents, failing—complex systems.[12]

A global assemblage approach serves to open up key concepts to the realities of the world rather than attempt to sew up neat and tidy understandings. Abrahamsen argues that

> when studying the global, the nation-state is generally approached as a unit that contains society, which in turn is separate from the unit of the international or the global. Assemblage thinking rejects such fixed and stable ontologies, and replaces essentialism and reification with a flat ontology.[13]

Increasingly the issues in global governance scholars are treating cannot be limited to territorially bounded entities and, as the gaps in the literature on CBNRM demonstrate, this proves equally to be the case with issues in global environmental governance.[14]

Assemblage thinking as an approach proves to be one useful way to overcome many of the ontological restraints that have limited previous studies. As the pace and complexity of global politics have increased, our vocabulary often lags behind.[15] For this project, concepts like state, society, legitimacy, and sovereignty still hold important analytical and descriptive value but an assemblage approach will be used to excise them from the straitjacket-like hold they usually maintain as foundational concepts in International Relations theory. Having been developed alongside the progression of international relations during a time when state-centricity made more analytical sense, these concepts largely concretized and resultantly failed to evolve alongside developments in world politics.

Abrahamsen posits "as a way of studying the global, assemblages are particularly attractive as they draw attention to new and emerging multiscalar connections, abandoning strict dichotomies between the global and the local, the international and the domestic."[16] For assemblage thinking

[12] Jane Bennett, "The Agency of Assemblages and the North American Blackout," *Public Culture* 17, no. 3 (September 1, 2005): 445–66.

[13] Rita Abrahamsen, "Security Privatisation, the State, and Development in the Global South," in *The Palgrave Handbook of International Development*, ed. Jean Grugel and Daniel Hammett (London: Palgrave Macmillan UK, 2016), 2.

[14] Elden, "Missing the Point."

[15] Rita Abrahamsen, "Assemblages," Routledge Handbook of International Political Sociology, December 1, 2016.

[16] Abrahamsen, 4.

what is at the centre of analysis is relationships between elements, rather than the elements themselves. This is important in enabling scholars to focus on provisionally and historically contingent relations between both human and non-human elements alike, rather than totalities and reified units of analysis.[17] Building on this Paul Rabinow argues, an assemblage "is not a pre-existing thing of the world with pre-given properties," but instead "brings together entities in the world into a proximity in which they establish relations between and among themselves while remaining external to each other and thereby retaining their original properties."[18] Assemblage thinking allows us to unstack the usual vertical approach that sees the global on top of the national, on top of the local in a hierarchical dynamic that artificially divides each realm.[19] As has been pointed to, this level of analysis approach ceases to provide analytical clarity to complex phenomena such as CBNRM that rearticulates power and those whose relations produce it.

Theoretically a flat ontology would suggest that all subjects or objects have the same degree of being-ness and that none of these units of analysis is superior or more central or more important than any others. Methodologically this would suggest that one enters the field not knowing what they expect to find—however, this of course brings with it a certain degree of methodological challenges as well. In order to prevent some of these challenges I approached this study with a flat*tened* ontology, yet not one that is entirely flat which would be both impractical in a number of ways and likely impossible as we know there is no view from nowhere. Approaching an issue such as CBNRM with a flattened ontology precludes some of the pre-determined assumptions that come from many theories that bring the baggage of state-centricity with them.

A flattened ontology allows us to engage with the subject and those related to it in order to find what is there and analyse how it operates, rather than seeking to find specific outcomes and reify existing approaches. This reification can serve to depoliticize issues and mask power imbalances that gloss over the ways in which certain endeavours such as wildlife conservation which is so often presented as an unquestioned 'good' but which might actually serve the interest of those most powerful, while either

[17] Acuto and Curtis, "Assemblage Thinking and International Relations."

[18] Paul Rabinow, *Anthropos today: reflections on modern equipment*, In-formation series (Princeton, NJ, Princeton, N.J.: Princeton University Press, 2003), 206.

[19] Abrahamsen and Williams, *Security beyond the State*.

failing to benefit marginalized populations or, in some cases, actually further marginalizing them or leading to them incurring costs associated with the conservation of certain wildlife species. Thus, despite methodological challenges and the complexities of a global assemblage approach and the associated flattened ontology, the approach also brings important opportunities to change perspectives and question established wisdoms and understandings. The notion of a flattened ontology also builds on work by Foucault who argued for methodological approaches that do not focus on pre-given universals but instead start from a position of examining how these categories are assembled through concrete practices—which is what I do with the empirical chapters of this book.[20] Ignoring or glossing over some of these more difficult and challenging issues of power and politics is a particular issue that Siba Grovogui has written about at length, arguing that scholars must move beyond Eurocentrism and traditional realist concepts such as anarchy in order to engage in a new International Relations that more adequately accounts for the modern world, and specifically for Africa which has so long been unknown by the discipline.[21] An assemblage approach attempts to accomplish this substantial task by not assuming a specific set of actors or distribution of power and authority, but instead sets out to discover this empirically.

I therefore analyse CBNRM as a global assemblage in order to trace the policy development and evolution as a result of a diverse number of component parts that were strongly informed by transnational flows of policy actors, ideas, and practices, which have manifested in place-specific ways. Achille Mbembe argues that "in order to account for both the mind-set and the effectiveness of postcolonial relations of power, we need to go beyond the binary categories used in standard interpretations of dominations, such as resistance v. passivity, autonomy v. subjections, state v. civil society, hegemony v. counter-hegemony, totalization v. detotalisation. These oppositions are not helpful; rather they cloud our understanding of postcolonial relations."[22] Assemblage approaches are specifically employed

[20] Foucault, Society Must Be Defended.

[21] S. Grovogui, *Beyond Eurocentrism and Anarchy: Memories of International Order and Institutions* (Springer, 2016); Luc Sindjoun, "L'Afrique Dans La Science Des Relations Internationales: Notes Introductives et Provisoires Pour Une Sociologie de La Connaissance Internationaliste," *African Sociological Review/Revue Africaine de Sociologie* 3, no. 2 (1999): 142–67.

[22] Achille Mbembe, "Provisional Notes on the Postcolony," *Africa* 62, no. 1 (January 1, 1992): 3.

to move beyond such binaries in order to unpack and analyse the complex power relations that exist in such postcolonial settings specifically in sub-Saharan Africa where state-centric theories fall particularly short.

With CBNRM we see the way policy ideas, practices, and forms of influence are strongly informed by transnational flows such as the desire of some to curb trophy hunting, omit climate change from the discussion, or seek to improve the sustainability of photographic tourism. However, the conditions of possibility for such policies depend largely on local conditions and actors to enable them. Thus, while the CBNRM policy seeks to engender certain effects, the extent to which it does or does not can perhaps only rarely be attributed solely to the policy itself. This is the way assemblages are characterized by relations of exteriority in which a component part of an assemblage may be detached from it and plugged into a different assemblage in which its interactions are different. Therefore, rather than relations between components parts being understood as logically *necessary* to make the whole what it is, relations in an assemblage are seen as only contingently obligatory.[23] An example that is given by Savage is a bicycle for which the component parts including chain, gears, pedal, and so on serve a strategic function in ways that are logically necessary to make the whole what it is. Relations in a policy assemblage also serve strategic functions but comprise contingently obligatory relations that are always evolving in form. In this sense a given NGO such as WWF serves a specific role globally as a leader in wildlife protection, however its various sub-branches which exist in many countries each serve specific but unique functions in each setting as they are individually influenced by the realities of the components and settings around them. Thus, WWF does not only exist as an actor within Namibia or in relation to the country's CBNRM programme but exists simultaneously beyond that specific assemblage as well as within it, leading to various and distinct effects in each setting similarly to the way the tires on a bicycle might serve similar and quite different functions in diverse settings.

By approaching CBNRM with a flattened ontology, I first and foremost do not engage with the subject assuming it is a policy that is passed by the Namibian state which then flows down to its subjects as is often traditionally understood of many policies. A key feature of the flattened ontology is not assuming pre-determined actors or structures, with the state being key among these. Amy Niang argues that

[23] Bennett, "The Agency of Assemblages and the North American Blackout."

> there is nothing natural, linear, convention or intrinsically consensual about the centralized state form. In fact, the African state emerged, and was erected against, and at the expense of a variety of authority structures and forms of self governance. The state has sustained itself through destructive practices, internal colonization, and in fact the production and alienation of a range of internal others.[24]

This book takes the state as being an ongoing and varied process rather than a pre-determined central construct from which policy and power emanate. Instead, the flattened ontology sees the research trace the developments of CBNRM policy through historical analysis, and then builds through to discourse analysis of secondary sources, grey literature, and interviews in order to piece together the heterogeneous and often disparate actors whose efforts and interests combine in cooperative and often simultaneously competitive relations. These relations resulted in the unofficial genesis of CBNRM prior to independence, which was eventually codified in the 1996 Nature Conservation Amendment Act and over the past 25 years has continued to evolve in response to challenges and events. However, the legislation does not keep evolving in the halls of government and instead is a result of the social relations of the actors that are traced in this inquiry.

While actor-network theory, political ecology, transnational networks, and other theoretical approaches can each do parts of this and have contributed to the further development of global assemblages as an approach, each does only part of what the assemblage approach does in this book. CBNRM is not a domestic policy for conservation that is then influenced by global power relations, but instead it is actually a policy assemblage that emerges through a confluence of global and local, public and private social relations and it cannot be understood outside of these and is specifically an example of global environmental governance at work. As discussed above, assemblage thinking is more than a traditional theoretical framework, but is also a methodological approach for engaging with these diverse units of analysis as the next section will further explore.

[24] Amy Niang, The Postcolonial African State in Transition: Stateness and Modes of Sovereignty (Rowman & Littlefield, 2018), 6.

Methodology

My method of inquiry began as more a strategy than a plan as assemblage work aims to move beyond some of the reification implicit with traditional theories, methods, and approaches. This strategy was partly to enable me to avoid the type of reifying approaches that assemblage thinking attempts to move beyond by not entering with fixed ontological assumptions 'knowing' which actors will be of particular importance and what forms of power they might exert.[25] I set out to study social relations and observe how they enact CBNRM in Namibia. As argued by Vrasti,

> the answers to our research questions rarely await us in the field. Often we return home more confused than we were in the first place. This is where ethnography comes in handy: [...] ethnography can afford to travel back and forth between the part and the whole, experience the text fieldwork and theory, certainty and epiphany in ways that other methods cannot and which, in the end, can only add to the credibility and authenticity of this genre.[26]

Explained this way, ethnography in many ways mirrors the methodological aspects of assemblage thinking and is part of what allows for the accumulation of data while attempting to avoid some of the reification that comes with the fixed ontological assumptions of other methods and theories. While the timeline for my fieldwork limited the depth of this data collection that was possible, I utilized traditional social science methods that mirror some of the key parts of ethnography in order to study the components of the subject matter in a synergistic approach with other methods of observation in order to triangulate data and draw useful conclusions.

This book examines CBNRM in Namibia in order to analyse its evolution in response to climate change. Namibia represents the crucial case for the study of CBNRM given its recently drafted and progressive constitution, its abundance of wildlife, and sparse human populations. It has also transferred more rights through the CBNRM programme to communities than any other country.[27] As indicated in Chap. 2, Namibia has also seen a

[25] Acuto and Curtis, "Assemblage Thinking and International Relations."

[26] Wanda Vrasti, "Travelling with Ethnography," Research Methods in Critical Security Studies, April 3, 2013, 61.

[27] Bollig and Menestrey Schwieger, "Fragmentation, Cooperation and Power."

higher percentage of its land devoted to CBNRM than any other country in the world.[28] Its programme is also praised for being more locally driven than those of neighbouring countries as well as having more complete devolution—broadly accepted within the literature as a key component for success.[29] As a result, Namibia has been touted by the international community as one of the greatest successes of CBNRM and as an example to be emulated and repeated elsewhere.[30] Crucial case studies are based on most likely or least likely designs based on the assumption that some cases can have more or less value for testing theories than others. For least likely case design it is based on what Levy refers to as the 'Sinatra inference'—being that if the case can work in this least likely scenario, it can work anywhere.[31] For this book I will employ the inverse of this with a most likely case; the logic being if CBNRM does not accomplish its purported goals even in Namibia it likely will not anywhere.

A crucial case study analysis is optimal for this book because of the scope and complexity of the environmental assemblage that enables CBNRM which will make this single case study of relatively *more* value rather than several cases. In utilizing the crucial case of Namibia, my research took me well beyond the borders of Namibia as the book posits that CBNRM is inherently globally oriented. This is seen by the web of NGOs, international financial institutions, and foreign governments that provide resources for the projects. But it is also seen through the assemblage of environmental activists, conservationists, and ecotourists whose activities and agencies influence environmental governance on the ground in Namibia in numerous and novel ways. By attempting to narrow the book's focus with a single case study, it will actually be able to provide a more complete analysis to unpack the depth and breadth of the global environmental assemblage. While I argue that due to their complexity, reach, and ever-changing nature, the entirety of any assemblage is likely impossible to unveil, by looking more deeply at a single case study, a more accurate and useful analysis of CBNRM will be developed than has been offered to date.

[28] Roe, Nelson, and Sandbrook, Community Management of Natural Resources in Africa.

[29] Boudreaux, "A New Call of the Wild."

[30] Hoole, "Place-Power-Prognosis."

[31] Jack S. Levy, "Case Studies: Types, Designs, and Logics of Inference," *Conflict Management and Peace Science* 25, no. 1 (February 2008): 1–18.

Part of Namibia's relative success with regard to CBNRM is attributable to its late independence—won only in 1990—which resulted in it having one of the most recently drafted and progressive constitutions in the world.[32] This constitution has environmental conservation written directly into it which creates a legislative framework through which to effectively devolve power over natural resources to communities.[33] Namibia was able to learn from neighbouring countries in the drafting of legislation in order to attempt to ensure its best chance at success.[34] These realities make Namibia an interesting and analytically useful case for manifold reasons. The international community and the broad network of NGOs and INGOs have had a great deal of influence on CBNRM from drafting legislation to project implementation, and ongoing management. This assemblage has yet to be adequately analysed and it has important effects in relation to developments in global environmental governance and hence the way power is reconfigured and rearticulated.

Namibia was also chosen as it is one of the countries that has contributed the least to climate change yet is one of those being hardest hit by its deleterious effects. With a population of approximately only 2.5 million it is a small country with many of its people remaining relatively poor, and as a result both historically and presently contributes fewer GHG emissions than the global average measured on both gross and per capita bases. Despite this, Namibia is expected to see substantial decreases in precipitation in the coming decades as well as an increase in mean temperature of 3 degree Celsius—well above the global average.[35] These climactic changes are wreaking havoc on Namibia and Namibians in general and in particular on the CBNRM programme as it relies directly on healthy and stable environments for success. The harmful impacts of climate change, which are extremely pronounced in the harsh conditions of Namibia, are directly correlated to benefits that can be derived from the programme, and hence the support it receives both locally and globally. CBNRM is a central component of the lives of so many Namibians that variations in programme implementation have a sizeable and measurable impact on the population

[32] Stephen W. Snively, "Environmental Laws of Namibia: Constitutions, Conservation, and Cheetahs," *Probate & Property* 26, no. 3 (2012): 44.

[33] Snively.

[34] Schnegg and Kiaka, "Subsidized Elephants."

[35] Kevin Coldrey and Jane Turpie, "Climate Change Risk Assessment for Namibia's Communal Conservancies," 2020.

and on the economy.[36] Similarly, policy shifts in one area or targeted investments in individual projects can produce results that further policy developments can build on. Therefore, analysing a country that is being affected in the same way much of the region will be in the coming years can prove a useful case study from which more generalizable conclusions can be drawn for other countries and studies to learn from. While research has shown that all regions of the world will be increasingly impacted by climate change, understanding these more substantial and deleterious effects in Namibia can help us develop solutions that might, in time, be exportable to other countries. This can be particularly useful for CBNRM programmes in other countries across Southern Africa but can also prove useful for other policies relating to sustainable development both on the African continent and elsewhere. Understanding these environmental impacts in Namibia, which will soon be impacting neighbouring countries throughout Southern Africa, is a key contribution of this book and a central reason why Namibia is a particularly informative case study for this analysis.

Within Namibia I have two main foci. First, I analyse CBNRM in the country as a whole, studying the policies put forth by the national government and unpacking what these mean for the country, how they are implemented, and how these interact with the international community. Second, at the community level my research focuses on conservancies in the Kunene region. The Kunene region is home to 35 conservancies—the most of any of Namibia's 13 political regions—as well as many of the oldest conservancies and lodges and some of the newest.[37] The region is rich in wildlife and also has one of the more diverse geographies and environments of the country. There is also useful existing research on CBNRM in Kunene, which allowed me to develop thicker analyses by building on prior work.[38] While I could have chosen a region that has been studied less in an effort to maximize originality, I see the work that has already been done as useful and unnecessary to duplicate. The existing work on the Kunene region forms the framework from which I begin to put together

[36] Coldrey and Turpie.

[37] Brian T. B. Jones, "Policy Lessons from the Evolution of a Community-Based Approach to Wildlife Management, Kunene Region, Namibia," *Journal of International Development* 11, no. 2 (1999): 295–304.

[38] Brian Jones, "The Evolution of a Community-Based Approach to Wildlife Management at Kunene, Namibia," *African Wildlife and Livelihoods: The Promise and Performance of Community Conservation*, 2001, 160–76; Bollig, "Towards an Arid Eden?"

the pieces of CBNRM to better understand how climate change has impacted the programmes and how actors and their activities are evolving in response to these constraints.

A note to add on climate change is that it did not factor into my initial proposal or book/research plan which was more centred around the politics and power relations involved in CBNRM. However, once the fieldwork began virtually every interview quickly led to discussions of climate change and the drought that was being experienced without me asking questions about it. This was a situation that continued throughout the data collection and something I distinctly witnessed throughout my experiences on the ground. As a result, the research itself pointed me to the centrality of climate change and the many ways in which actors were at work contributing to evolutions of the programme in response to the increasing challenges brought on by the changing environment. This also demonstrates one of the ways in which global assemblages is an effective methodology as the flattened ontological approach led me to what awaited in the field rather than finding what I expected there to be.

Assembling Methods of Observation

Methods of observation included semi-structured key informant and elite interviews, focus groups, participant observation, and multi-level policy and documentary analysis. In addition, I examined secondary sources as well as grey literature including data on finances, wildlife, and governance from conservancies, policy documents, and conservancy reports. Combining these methodological approaches is effective as focus groups risk having certain voices heard above others yet allow for the observation of a number of voices coming together to formulate a consensus, or to demonstrate disagreement.[39] Participant observation enabled me to study various groups interacting and the types of power that can be exhibited with as little of my interaction as possible.[40] However, neither of these approaches provided the richness and depth of discussion as semi-structured interviews.[41] As a result, each of these approaches has several

[39] Jenny Kitzinger, "The Methodology of Focus Groups: The Importance of Interaction between Research Participants," *Sociology of Health & Illness* 16, no. 1 (1994): 103–21.

[40] Susan Kang, "What the Documents Can't Tell You: Participant Observation in International Relations," *PS: Political Science & Politics* 50, no. 1 (January 2017): 121–25.

[41] Nicholas Clifford et al., *Key Methods in Geography* (SAGE, 2016).

strengths as well as certain limitations. Layering each together allowed me to triangulate a wider breadth of data to draw more substantive conclusions than would be possible with a narrower approach.

My fieldwork yielded 75 semi-structured formal or informal interviews,[42] 4 informal focus group discussions, and several opportunities for participant observation in a range of diverse settings. I spoke with government officials ranging from local-level officers to the Minster of Environment and Tourism, staff from a variety of domestic and international NGOs, academics, consultants, tourism companies, tourism company employees and managers, farmers, mine workers, rural villagers with diverse livelihoods, as well as conservancy staff, elected officials, game guards, and police. In order to obtain a high number of carefully selected participants for interviews I used a snowball approach that built on my initial more formal process of recruitment.[43] These formal means began prior to travelling to Namibia and included emailing and calling a number of relevant contacts produced from the websites of government ministries, post-secondary institutions, NGOs, and tourism enterprises. One of the most useful of these websites was the Namibian Association of CBNRM Support Organizations (NACSO) which is an umbrella organization that facilitates coordination between the network of NGOs and other actors that contribute to CBNRM in Namibia. The NACSO website has a list of all its major partners which are key actors for CBNRM in Namibia, and this provided me with a useful base of contacts that I developed in the months leading up to departure.[44]

Often these formal avenues helped secure interviews with high-level government and NGO officials which were especially useful in gaining insight into decision making, policy formulation, and broad overviews of the programme. They also provided valuable insight about the influence of international actors as well as major challenges and benefits these partnerships provide to the programme. These more formally procured participants were also pivotal in 'getting the snowball rolling' as these respondents often provided me with a list of contacts or further avenues to pursue. From there, I was able to continue to successfully find carefully

[42] A full list of interviews can be found in Appendix A.

[43] Julian Kirchherr and Katrina Charles, "Enhancing the Sample Diversity of Snowball Samples: Recommendations from a Research Project on Anti-Dam Movements in Southeast Asia," *PLOS ONE* 13, no. 8 (August 22, 2018): e0201710.

[44] NACSO, "Members," NACSO, 2015.

selected contacts. This avenue led me down a path which could yield a biased sample—those who already support CBNRM and have a stake in its perceived success. To mitigate such biases, I balanced suggested interviews with contacts made on my own through a variety of methods, including cold calling people over the phone, emailing, as well as knocking on doors and meeting people in the streets and villages. In engaging with respondents that had been suggested to me I also asked questions and analysed the responses understanding that they are inserted in a particular way into the politics of CBNRM and would be responding as such.[45]

The interlocutors contacted through formal methods offered rich and knowledgeable insight into the details of the programme and form an important base of my research. However, those I interviewed as a result of snowballing provided a more granular level of detail than the high-level interviews—the kind which is much more difficult, or in many cases impossible, to find online or in existing literature. Beyond this, employing the snowball method proved a critical part of my data collection as it demonstrated the social and political makeup of CBNRM in Namibia. This provided invaluable experience to analyse various dynamics in order to see how these actors engage in decision making, resolve differences, cooperate, and in many cases gain first-hand insight into how they feel about one another in less formal settings. These relationships are understudied and are often presented as working more or less in harmony towards congruent goals of conservation and sustainable development by some.[46] While the goals of these actors are often analogous, there remain important political dynamics and power struggles that have meaningful effects on outcomes. Mapping these diverse actors through both formal and informal processes at a number of distinct levels provided key insight into the politics of CBNRM beyond simply a policy for domestic conservation. While the list of interviews, which can be found in Annex I, might leave some wondering how 'global' this assemblage in fact is, as many of those interviewed might seem more 'local,' assemblage thinking does away with this type of strict differentiation and place becomes less important for understanding various types of actors. An assemblage approach does away with the levels of analysis and hierarchical understandings of the global on

[45] Kirchherr and Charles, "Enhancing the Sample Diversity of Snowball Samples."

[46] Brian Child, "Origins and Efficacy of Modern Community Based Natural Resources Management (CBNRM) Practices in the Southern African Region," *Snively*, 2003; Snively, "Environmental Laws of Namibia."

top of the national which is on top of the local and which generally understands the UN and World Bank as being key global actors, while an NGO operating in Namibia's Kunene region as a local actor. However, place becomes less important than connections with the assemblage and as a result my ontological approach sees even local seeming actors as inherently global in the way they are connected to broader power relations and actors around the globe. As a result, I allowed the research to lead me down a path that was not pre-determined and I did not go searching for specific actors whether they be 'global' or 'local.'

The research strategy employed led me to a number of interconnected sites which allowed me to draw useful data. This began in the capital, Windhoek, where many government offices and NGO headquarters are located. One of the most informative of these informal connections was an invitation I received to a full day workshop with the NACSO members as well as a number of other affiliates and stakeholders. This workshop supplied a wealth of information as well as a useful opportunity to analyse some of the high-level politics and deliberation that goes on between key decision-making actors in CBNRM.

Following this stage of the research, heading to the Kunene region there were many people, places, and projects I was suggested to visit which led me to the ten key conservancies studied in this book: Tsiseb, Sorris Sorris, Doro !Nawas, Uisbasen/Twyfelfontein, Torra, Palmwag, Sesfontein, Anabeb, Purros, and #Khoadi- //Hoas. Spreading the research across ten conservancies inevitably constituted a trade-off between depth vis-a-vis breadth of knowledge, and this was a decision I consciously made in order to study potential variations within the region. Each conservancy supplied an interesting and important piece of the puzzle unpacking CBNRM and the complex politics of the assemblage. It also continuously demonstrated the global involvement at the local level that is often difficult to ascertain from the literature alone. Here the snowballing technique led to the unveiling ceremony of a new solar plant at the Sorris Sorris Lodge in Sorris Sorris Conservancy which proved to be perhaps the single most useful snapshot of the actors that contribute to the layered governance assemblage that is CBNRM. The ceremony was a large and elegant affair with a number of officials from multiple levels of government including the Minster of Environment and Tourism. There were also traditional authorities, elected conservancy leaders, NGOs, and representatives from key international organizations which had partnered with the conservancy to make the solar plant possible.

The solar plant unveiling included a full day of presentations, speeches, meetings, discussions, meals, and performances from local children's choirs, as well as a large red ribbon cutting—all of which elucidated the importance with which this major development was viewed. At this event I was introduced to the manager of one of the nearby conservancies I had planned to travel to soon after and she informed me that I would be there during the annual audit of their conservancy which I was welcome to observe. This again proved to be a rich source of data as I not only camped with the auditing team that travels to conservancies to perform audits which allowed me to interview a number of its members, but then also spent the next day observing the auditing process. The audit team is composed of members of the government as well as both domestic and international NGO staff in order to ensure reporting is fair and accurate and thus this proved another key piece to my research puzzle.

Having made the connections I did in Windhoek through my initial formal approach, these interview respondents helped me develop key contacts along the way and thus when I would arrive at each conservancy they would be expecting me and proved extremely open and helpful with the research. In some conservancies this consisted of an hour-long conversation with the conservancy manager or chairperson, and in others it was a meeting lasting the better part of a day with several members of the conservancy council, traditional authorities, and local conservancy members. These were very useful in hearing from a number of voices from diverse backgrounds to provide rich information to begin to effectively triangulate the data in order to draw meaningful conclusions. Following these initial meetings, I would receive blessing from either conservancy leaders or traditional authorities to branch out on my own—or with an interpreter in one community—and begin exploring local communities and villages to approach various people and request interviews. Everyone I approached was open to speaking to me once they found out who I was and what I was there for. This informal snowball technique proved useful for gaining rich insight from a variety of Namibians who were not specifically tied to any organization and resultantly were likely to speak more freely and honestly as they did not stand to benefit or incur punishment as a result of particularly positive or negative outlooks on the programme.

In Namibia there are a number of languages and dialects that are predominant in various regions across the country; however, the official language is English and, as such, the majority of people have at least some level of English competencies.[47] It is possible that those who do not speak English may have been less willing to agree to be respondents; however, I approached a wide variety of people and they were generally open to responding and sufficiently able to converse in English. While this may have excluded certain voices, the relatively large number of respondents would be likely to minimize potential sample biases. English competencies are also continuously improving as CBNRM has developed and as tourism continues to be a growing part of the Namibian economy.[48] Only in Anabeb Conservancy was there a generally lower level of English spoken and as such Godlieb, the conservancy office manager, offered his services to join me following the official morning meeting with the conservancy council and traditional authorities to act as an interpreter in the community. For this part of the research, I chose who to speak to and each respondent spoke at least a minimal level of English so they generally had an idea of what I was asking and at times would offer responses in English. Thus, while it is possible that having the conservancy office manager present and acting as an interpreter could have led to certain biases in the reporting of responses, these were generally mitigated and only in one of the ten conservancies I was active in.

With the nature of this research, it is integral to explicitly engage with the fact that there are inherent power differentials built into the research between interviewees and the interviewer. In their most basic form, these consist of me being a white, male researcher from the Global North, with a certain level of post-secondary education as well as a particular socioeconomic background. While there is no way to eliminate these imbalances which can lead to biases in research samples and data collection, acknowledging these is a necessary starting point in order to mitigate their impact on the research. Assemblage thinking as methodology serves to assist in this endeavour as it attempts to approach complex phenomenon by looking beyond traditional power structures and hierarchies in order to disaggregate the types of structures that exist in society. While accomplishing this is likely an impossible endeavour, it is necessary for researchers to enter the field knowing they are positioned in a certain way vis-a-vis their units of analysis and develop approaches to account for this in their actions. This is something I worked to accomplish each day and through each

[47] Godfrey Mwakikagile, *Namibia: Conquest to Independence: Formation of a Nation*, 1st edition (Dar es Salaam: New Africa Press, 2015).

[48] Mwakikagile.

interview, by selecting a diverse pool of interview respondents, and by adhering to the principles of the Tri-Council Policy Statement: Ethical Conduct for Research Involving Humans. To further maximize the likelihood of receiving honest answers and mitigating potential biases, I have kept all rural respondents anonymous and informed them I would do so in gaining their consent to participate in the interviews. This option was chosen as interviews were based on their lived experiences and their specific titles provided no further weight to their interviews in the way a government official or NGO director would. I also chose this approach to help ensure that they would not fear backlash from the community or government for speaking ill about the programmes or leadership. Alternatively, I did not want them to feel incentivized to speak particularly positive about the programme in hopes of being rewarded should this news get back to someone in a leadership position.

In order to analyse the data collected I employed both historical and discourse analysis as a method for unpacking various actors' understandings of CBNRM within their specific sociopolitical and institutional contexts. This method of data analysis is effective for understanding data that stems from interviews as it excavates ways in which interview respondents rely on a specific framing of the issue at hand as well as the complex and multiple social outcomes of that framing. Therefore, I examined explicit and implicit patterns and regularities that emerged from their understandings as well as how various functions of CBNRM are practised across the research sites. Thus, I closely analysed responses to see where various actors seemed to share similar talking points, where they agreed, where and in what ways they disagreed, as well as ways in which their interpretations might be conflictual at times, even when overtly they display relations of cooperation and harmony. In order to triangulate data, I balanced grey and secondary literature, with interviews, and personal experiences/participant observation. My own experience of driving through dry riverbeds, or seeing many riverbeds dry throughout my fieldwork, is of course not sufficient evidence in itself, therefore this was triangulated with adding in discussion of the fact that I was there throughout the rainy season and that the many people I spoke with said the rivers had been dry for years. With the interviews leveraged I wanted both to indicate the experience and perceptions of the people in the conservancies, as well as to analyse the manner in which CBNRM adapted to climate change. In other words, there is triangulation of evidence when I seek to establish who influences the policy, but also an attempt to account for how the programme is perceived and experienced by the target population.

Finally for this chapter, a note on the limitations of assemblage thinking specifically with respect to generalizability. As discussed, Namibia is the driest country in sub-Saharan Africa, and it is one of the countries being hardest hit by climate change. While we know that climate change does not impact all countries and regions the same or even in the same ways, we do know that the entire world is already experiencing impacts of climate change and that these will only be exacerbated in the years to come. Furthermore, as CBNRM is a form of resource governance that relies specifically on healthy and functioning ecosystems and natural resources, a changing climate will impact CBNRM activities and its viability everywhere. While we cannot say precisely how the rest of Southern Africa will be impacted by climate change in comparison to Namibia, research has clearly demonstrated the rest of the region will similarly be impacted by rising temperatures, along with ever-decreasing precipitation and increasingly erratic rainfall patterns.[49] Meaning, to some degree or another the rest of Southern Africa will face similar worsening environmental and ecological challenges, and thus, so will their CBNRM programmes, as well as other sustainable development initiatives that rely on healthy and functioning ecosystems—which is basically all of them. This is demonstrated by research from Pierre vas Rensburg and Cecilia Tortajada who illustrate that

> like in other parts of the world, there is a substantial increase in concurrent droughts and heat waves in Southern Africa. The 2015–2016 flash drought (droughts with a rapid onset and short duration but with high intensity and devastating impacts) that affected the region was characterized by severe heat waves and soil moisture deficit, the highest since 1948. Flash droughts have increased by 220% from 1961 to 2016, and they are likely to further increase in the future.[50]

The NASA Earth Observatory considered this event to have been unprecedented as the combination of decreased and delayed precipitation as well as increasingly rising temperatures threatened the food security,

[49] NASA Earth Observatory, "Drought Threatens Millions in Southern Africa," Text. Article (NASA Earth Observatory, December 16, 2019).

[50] Pierre van Rensburg and Cecilia Tortajada, "An Assessment of the 2015–2017 Drought in Windhoek," *Frontiers in Climate* 3 (2021): 1.

fresh water supply, and energy needs of millions of people.[51] For the city of Cape Town the drought amounted to what was considered a once in 590-year event based on historical rainfall records. The impact of this event on people and local and national economies was severe and exacerbated by a failure to adequately plan for and manage such an event.[52] As a result, if we can understand these challenges in one region, as well as their impacts and ways in which they are both successfully and less successfully overcome, we can learn from this and generalize across other countries and programmes.

That being said, with an assemblage approach being empirically informed, generalizations, as laid out above, are possible, but specificity remains an issue as each really seeks to excavate the depth and breadth of the specificity of any context which is itself always changing and will always be different from any other. The ontological approach suggests that CBNRM policy in Namibia might contain certain components that are arranged in particular ways and operate with positive impacts in that context. However, this does not mean these same components will have similar positive effects in Bostwana or Zimbabwe for example, and instead might have different positive impacts or even produce negative outcomes depending on the local contexts. This is because when the components are assembled in a new context, the components themselves will be controlled by numerous context-dependent factors or conditions of possibility which will render the components place specific and result in new relations being established and maintained between these components and existing components in the new environment. The current study and the implications for its generalizability are for several reasons limited to Southern Africa specifically. Within Southern Africa, there are indeed a number of forms of CBNRM; however, the vast majority of CBNRM programmes across the continent consist of some form of wildlife conservation for the purposes of trophy hunting or ecotourism, similar to what is described in this book. Of course, there is variability, and this study will mean less to a CBNRM programme in Tanzania based on community fisheries for example, but these programmes are much more rare, and as such the book aims to understand the predominant form of CBNRM and to build generalizability for the region for that specific approach. Nonetheless there are clear limitations built into this approach as it depends crucially on how societies are

[51] NASA Earth Observatory, "Drought Threatens Millions in Southern Africa."

[52] City of Cape Town, "Our Shared Water Future. Cape Town's Water Strategy," 2019.

organized, as well as the various actors involved and the nature of the state. The assemblage approach is developed to theorize from the ground up and thus while I do so with an eye to generalizability, over-generalizing undermines the notion of a flattened ontology and returns to assuming what actors are involved and how power relations operate between them. Thus, each case can build from mine however each would require its own nuance and some degree of empirical findings of its own. The next chapter will dig more deeply into the specific context surrounding the emergence of Namibia's CBNRM based on its particular history.

CHAPTER 4

The Local Genesis of Global Environmental Governance in Namibia

Next the book turns to the empirical chapters with this chapter outlining the genesis of CBNRM in Namibia which will include a brief history from the colonial period, through to the Apartheid era, and onto independence in 1990. This history will include key relevant background on the politics of the country including its de facto one-party state system that has emerged since independence. Through the genesis of CBNRM, this chapter begins unpacking the ways in which this form of resource governance is about much more than a straightforward devolution of state power to local communities. Beyond this, the chapter discusses questions of local-level corruption which are key to understanding the degree to which communities benefit from CBNRM. This form of resource governance was intended to circumvent the type of neopatrimonial practices and large-scale corruption that have often been seen at the state level in Africa.

> *It is both futile and an insult to the poor to tell them that they must remain in poverty to protect the environment.*[1]

This chapter discusses the genesis of CBNRM in Namibia and the historical context that led to the policy as well as how this history impacts

[1] World Commission on Environment and Development, *Our Common Future* (Oxford; New York: Oxford University Press, 1987).

A. Heffernan, *The Global Politics of Local Conservation*, Environmental Politics and Theory,
https://doi.org/10.1007/978-3-031-24177-2_4

ongoing resource governance in the country. Understanding this history is key in order to analyse the evolution of the CBNRM policy and programmes over time. Part of this history begins with the need to briefly explore Namibia's colonial occupation and subsequent oppression under South Africa's Apartheid regime. Together these governance structures exploited Namibia and its black populations for over a century which created harmful and lasting power imbalances that remain meaningful and evident today. Independence was won only in 1990 after a protracted and violent struggle which set the stage for a new approach to land management and conservation in the country. However, conservation efforts and the shift to community-based management were also underway prior to independence. All of this is integral to understanding how the policy emerged as well as the assemblage of actors who contributed to the development of CBNRM and its ongoing evolution today.

The policy framework for CBNRM in Namibia was laid out with the 1996 Nature Conservation Amendment Act. This legislation enables communities to form governance entities known as conservancies, to which the central government would then devolve control over natural resources in order for them to be managed sustainably at the local level. This chapter discusses the formation of Namibia's assemblage by outlining the CBNRM policy formulation and the shifting assemblage that it has consisted of over time. It is important to analyse the actors who contributed to both the development and implementation of the policy in order to further unpack the politics of this form of resource governance in the subsequent chapters. Understanding the political manoeuvring that led to the development of CBNRM is key to analysing the evolution of the policy assemblage in response to climate change.

From Colonial Occupation to Apartheid Rule

The purpose of this section is not to offer a comprehensive political history of Namibia, but instead to provide the necessary historical background for understanding key issues related to CBNRM. Namibia has a colonial history that has had an ongoing impact on present day politics. The same can be said for many countries in sub-Saharan Africa but for Namibia, its recent independence—gained only in 1990—presented it with a particular set of opportunities as well as significant and ongoing challenges.[2] In order to understand CBNRM or any other recent political developments in the country it is essential to understand this history. The territory currently known as Namibia—like the rest of the African continent—has a long and rich history that pre-dates the colonial era and stretches back well before Europeans first began activities in the region. Although this history is important and should not be ignored, for the purposes of this study it is far too extensive to address at length. Suffice it to say that what we now call Namibia and Namibians have their own rich and diverse history that existed prior to colonial occupation.

Beginning in 1884 following the Berlin Conference which kicked off the 'Scramble for Africa,' German forces occupied what became German South West Africa, which I will henceforth refer to as Namibia (although it did not become that until 1990).[3] For the next 30 years German colonial forces solidified their presence in the area, building forts, extracting resources, and basing much of their power on the violent enforcement of oppressive colonial force much like across most of the rest of the continent. This included a genocide in the early twentieth century in which up to 100,000 Herero, 10,000 Nama, and an unknown number of the San people were slaughtered by German colonial forces.[4] Beyond the violence employed against human populations, colonial rulers in Namibia also established much of their regime around the control of wildlife.[5] Wildlife and land are inextricably linked, and Namibians past and present are similarly connected to both wildlife and the land on which it resides.[6] During the early colonial period, many Namibian societies were still largely nomadic hunter-gatherers and thus their livelihoods depended on access to wildlife populations for food, tools, clothing, and more. Many also depended on wildlife for these purposes on top of relying on various forms of farming—usually in the form of pastoralism and the raising of livestock on open communal land.[7] Thus, limiting access to wildlife or land was one of the most direct ways for white colonial authorities to exercise their will

[2] Snively, "Environmental Laws of Namibia."

[3] Muriel Evelyn Chamberlain, *The Scramble for Africa*, 3rd ed., Seminar Studies (Oxfordshire, England: Routledge, 2009).

[4] Mwakikagile, *Namibia*.

[5] Schnegg and Kiaka, "Subsidized Elephants."

[6] Interview, Uatirohange Tjiuoro, Ministry of Environment and Tourism: Directorate of National Parks and Wildlife, January 20, 2020.

[7] Interview, Eric Xaneb, Chairman, Tsiseb Conservancy, January 27, 2020.

over black African populations by undermining their power and means for sustenance.

Namibia's colonial legacy is important for underlining the degree to which colonial authorities attempted to subjugate African states in order to benefit from them and their natural resources. This process was demonstrated following Germany's losses in World War I, as the League of Nations commissioned the Union of South Africa to administer the territory under a Class C Mandate.[8] For Namibia, this amounted to an arbitrary shift in political control that began a new colonial period which would last another 40 years until the Union of South Africa collapsed towards the end of the 1950s. This then led to the formation of the Republic of South Africa which soon after began to establish its Apartheid rule.

> Under the C mandate of the League of Nations, which entrusted South Africa with the administration of South West Africa/Namibia on behalf of the British crown South Africa was allowed to extend its judicial system of South West Africa/Namibia, which it did until Namibia's independence. Inexorably, the extension of South Africa's legal system to Namibia was co-extensive with the introduction of apartheid laws and policies in Namibia in 1948.[9]

This was a period marked by brutal, racist, and oppressive laws and systems of governing which Mahmood Mamdani summarizes as "bifurcated power that mediated racial domination through tribally organized local authorities, reproducing racial identity in citizens and ethnic identity in subjects."[10] Apartheid and its legacy had nothing but extremely detrimental effects on Namibia and its people and these effects continue to be vividly evident today.[11] The impacts of this oppressive regime similarly have meaningful effects on how wildlife is managed, land is distributed, and the political economy of conservation plays out in the country. The 1996 Nature Conservation Amendment Act was drafted

[8] Schnegg and Kiaka, "Subsidized Elephants."

[9] Mwakikagile, *Namibia*, 88.

[10] Mahmood Mamdani, Citizen and Subject: Contemporary Africa and the Legacy of Late Colonialism, Citizen and Subject (Princeton University Press, 2018), 2.

[11] Brian Jones, "The Evolution of Namibia's Communal Conservancies," in *Community Rights, Conservation and Contested Land* (Routledge, 2010), 119–33.

> to amend the Nature Conservation Ordinance, 1975, so as to provide for an economically based system of sustainable management and utilisation of game in communal areas; to delete references to representative authorities and to provide for matters incidental thereto.[12]

The 1975 Nature Ordinance had been the previously established colonial environmental legislation that had been oppressive, ineffective, and racist.[13] This is demonstrated as, at independence

> privately owned freehold land, amounting to 48 per cent of the territory, remained in the hands of less than 5000 mainly white farmers, while over 70 per cent of the population remained directly or indirectly dependent upon the 35 per cent of available communal land (the remaining 17 per cent is state-owned and largely nature reserves).[14]

This left black populations with small and unproductive territories for communal farming, ensuring they were kept in a situation of poverty and void of power.[15] This legacy of racism build into land and resource rights is also one of the central reasons why policies aiming to correct such issues must be crafted with care in order to ensure land reform and biodiversity conservation are complementary rather than conflictual.[16]

Apartheid was more than anything about power, and in many ways, Apartheid rulers sought to empower themselves by undermining the relative power of those they controlled rather than by actually strengthening their own means.[17] This unfolded by continuing the colonial legacy that deprived basic rights and legal protections of black and coloured people, including rights to land and wildlife. However, these efforts also went further to actively restrict any ability to benefit from local natural resources which limited the capacity of local populations to develop effective

[12] MET, "Nature Conservation Amendment Act," 1996, 2.

[13] Richardson, "Wildlife Utilization and Biodiversity Conservation in Namibia."

[14] Henning Melber, "Colonialism, Land, Ethnicity, and Class: Namibia after the Second National Land Conference," *Africa Spectrum* 54, no. 1 (April 1, 2019): 73–86.

[15] Jessica Brown and Neil Bird, "Sustainable Natural Resource Management in Namibia," *Development Progress*, 2011.

[16] Thembela Kepe, Rachel Wynberg, and William Ellis, "Land Reform and Biodiversity Conservation in South Africa: Complementary or in Conflict?," *International Journal of Biodiversity Science & Management* 1, no. 1 (March 1, 2005): 3–16.

[17] Frederick Cooper, *Colonialism in Question: Theory, Knowledge, History* (Berkeley: University of California Press, 2005).

livelihoods. Furthermore, the Apartheid regime ensured that those they wished to disempower remained undereducated, undernourished, and underdeveloped—thereby limiting their abilities to effectively exercise any form of power and these effects continue to be felt today.[18]

Over time the international community came to view South Africa's Apartheid system for the racist and oppressive regime that it was and eventually came to condemn it. The UN ceased to recognize Namibia as a mandate of South Africa, but this was not enough to free Namibia from the shackles of Apartheid Rule. By 1968 the UN and the Organization for African Unity (OAU) formally announced that Southwest Africa would henceforth be known as Namibia, but it would only be after a long and bloody war that Namibia would finally win its freedom.[19]

From the beginning of colonial occupation and through to the present day, conservation and particularly wildlife have been intimately enmeshed in Namibian politics and particularly in the politics of race and gender. To accomplish their goals, first German colonial authorities and subsequently the Apartheid system enacted oppressive and draconian environmental regulations to harness conservation and extract natural resources to increase their power and wealth. The history of conservation in the country is complex. While often laws and decrees that were enacted in the name of conservation served to oppress and exclude local populations, they also in some cases served important conservation functions. Though despite including important conservation goals, they were nothing but harmful and oppressive to local human populations. While colonial regimes ultimately had detrimental impacts on wildlife populations, their creation of national parks also served as sanctuaries for many species to be protected.[20] One of the clearest examples of this is Etosha Park in northern Namibia which is Namibia's crown jewel in conservation. Etosha was first established in 1907 as an official game reserve by German colonial authority Dr F von Lindqvist who felt it was necessary to protect the area's charismatic megafauna—though certainly his and similar colonial motives were far from altruistic.[21] Today covering approximately 13 per cent of terrestrial Namibia (ranking ninth in the world among land protected

[18] Boudreaux, "A New Call of the Wild."

[19] Mwakikagile, *Namibia*.

[20] Interview, Willie Boonzaaier, Programme Director, Integrated Rural Development and Nature Conservation, January 24, 2020.

[21] "Etosha History," Etosha National Park, accessed February 29, 2020.

through parks), national park politics remain contentious, and these parks serve to attract largely wealthy, white, foreign tourists who travel to experience Namibia's wildlife populations first-hand.[22] Local residents however were also initially moved from the land designated for parks and continue to be largely kept from the land and from benefitting from its resources—including wildlife populations. As a result, local black populations are often precluded from either enjoying the beauty of the local parks and wildlife or developing livelihoods from the natural resources within their limits. This is one way colonial sovereign power has continued through the Namibian state which has maintained these fortress approaches to conservation in a coercive fashion. As discussed, however, increasingly during this time power was becoming dispersed and did not always reside within the state or at the centre. As Foucault writes, where there is power, there is always opposition,[23] and this opposition to the exclusion of land and natural resources is precisely one of the issues that arose beginning in the 1980s and which translated into the CBNRM programme following independence in the 1990s which the next section will outline.

New Opportunities for Local and Global Sustainable Development

During the 1980s the Communal Areas Management Programme for Indigenous Resources (CAMPFIRE) was developed in Zimbabwe and was eventually followed by similar programmes in Zambia, Botswana, and Tanzania. From these, Namibia began to borrow parts of policies and work with actors in formulating similar policy plans of its own.[24] This process in part demonstrates the transnational nature of these projects. While each CBNRM programme is unique to each country and set out by state legislation, the policies emerged through the actions and conversations of a multitude of actors spanning these countries. CBNRM as a specific type of approach that sought to link local populations with natural resources to

[22] Luregn Lenggenhager, Ruling Nature, Controlling People: Nature Conservation, Development and War in North-Eastern Namibia since the 1920s (African Books Collective, 2018).

[23] Michel Foucault, *Security, Territory, Population: Lectures at the Collège de France, 1977-78* (Houndmills, Basingstoke, Hampshire ; Palgrave Macmillan, 2007).

[24] Berkes, Kofinas, and Chapin, "Conservation, Community, and Livelihoods"; Fabricius and Collins, "Community-Based Natural Resource Management"; Leach, Mearns, and Scoones, "Environmental Entitlements."

have them coexist was also similar to the Parks and People programmes and Transfrontier Conservation Areas. These among many others are issues that Maano Ramutsindela has written about at length discussing the interconnectedness of these approaches and the necessity of people and natural resources including wildlife and habitat coexisting peacefully in order to realize sustainable future.[25] These and other similar types of conservation approaches all share similarities in governing wildlife as wildlife populations migrate, but people must observe borders which means that CBNRM—like the other approaches mentioned above—is inherently a regional policy issue which is one of the integral points in suggesting CBNRM is best analysed as a global assemblage rather than a domestic policy tool for conservation.

The 1980s, which saw a rise in these transnational approaches to conservation, were also marked in Namibia by the violent struggles for independence and this meant the proliferation of armed groups stationed throughout the country.[26] Often wildlife were innocent casualties of war, their habitat was wiped out from fighting, they were hunted to feed troops, or they were simply slaughtered to prevent black Namibians from feeding themselves and their fighting forces. Moreover, poaching was a growing issue as wildlife were illegally hunted for their parts to be sold on the black market.[27] There was also a severe and protracted drought through the first half of the 1980s.[28] These combined forces led to a steep decline of wildlife populations that had already been facing immense pressures since the beginning of the colonial era and the explosion of demand for animal products in Europe and Asia. It becomes important to break down the long policy development process to understand who the actors were within the region. There were various conservation activists, nascent NGOs, local governance structures headed by traditional authorities, as well as communities across the country who were noticing a drastic drop in wildlife populations.

In response, various actors began working independently to set up community-based conservation programmes throughout Namibia. As

[25] Ramutsindela, Transfrontier Conservation in Africa.

[26] Interview, Pohamba Shifeta, Minister of Environment and Tourism, January 29 2020.

[27] Interview, Uatirohange Tjiuoro, Ministry of Environment and Tourism: Directorate of National Parks and Wildlife, January 20, 2020.

[28] Karine Nuulimba and Julie J. Taylor, "25 Years of CBNRM in Namibia: A Retrospective on Accomplishments, Contestation and Contemporary Challenges," *Journal of Namibian Studies : History Politics Culture* 18 (December 16, 2015): 89–110.

these actions materialized from various places and actors with different goals and priorities there is no clear consensus on what the initial impetus was for the CBNRM programme. This also demonstrates ways in which we see power being exercised through a variety of practices which together can be said to form the state, rather than the state acting in a quasi-top-down fashion that contains society. These practices are key to understand in analysing CBNRM as a policy assemblage emerging from the independent actions of a number of actors, rather than as a policy decree from the state. Some interview respondents—especially those with purely conservation backgrounds and foci—suggest that CBNRM came about as a direct response to declining wildlife populations in the 1980s and for the inherent need to protect these.[29] Several interviewees, however, dispute this as they indicate they did not see the value in wildlife beyond the ability to put meat on the table to feed their families. They saw wildlife as a resource that they could benefit from economically, but only if they were managed sustainably. As a result, they began to realize conservation was important for economic purposes.[30] Others, such as government officials, international organizations, and NGOs that focus on issues of governance, suggest that CBNRM policy emerged as a way to empower local communities and to provide them with the legislative tools necessary in order to develop local democratic institutional and governance capacity.[31] This camp views wildlife and natural resources as the catalyst for CBNRM but sees the development of organizational capacity as the main impetus for the programme which could grow beyond wildlife and resource management into the future. Ultimately, these three goals through which CBNRM emerged, became the three pillars of CBNRM and are largely understood as the three main objectives sought by CBNRM programs.[32] It is this set of heterogeneous actors and their social interactions that led to the formation of a global assemblage that would begin to implement community-based conservation activities which continue today—albeit in a much different

[29] Interview, Michael Knight, Transboundary Leader for KAZA, WWF, January 23, 2020; Richard Diggle, Business/CBNRM Specialist, WWF, January 24, 2020; Dr Margaret Jacobson, co-founder IRDNC, co-director, trustee, mentor, Conservancy Safaris Namibia, Swakopmund, February 17, 2020.

[30] Interview, Joglinde, Manager, Uisbasen/Twyfelfontein Conservancy, January 30, 2020; Emil, Manager, Torra Conservancy, February 2, 2020.

[31] Interview, Uatirohange Tjiuoro, Ministry of Environment and Tourism: Directorate of National Parks and Wildlife, January 20, 2020.

[32] Heffernan, "Development, Conservation, Empowerment: The Trilemma of Community-Based Natural Resource Management in Namibia."

and ever-evolving form. The key takeaway here is that the actions of a diverse set of actors were underway working towards largely congruous goals prior to independence and CBNRM legislation.

What can be referred to as the earliest example of CBNRM projects—though they were not yet called this—began in the 1980s in the Zambezi region in the far Northeast of Namibia. These activities were the brainchild of Garth Owen-Smith, a local academic, activist, and environmentalist who was eventually awarded the Global Environmental Prize for his efforts in 1993 jointly with Dr Margaret Jacobson.[33] Owen-Smith sought a way to utilize the conservation of wildlife in order to benefit local populations and remedy some of the ongoing issues of inequality, oppression, and underdevelopment facing black Namibians. Dr Jacobson joined him in his endeavours, and they ultimately formed Integrated Rural Development and Nature Conservation (IRDNC)—one of Namibia's first conservation NGOs which, today with a team of 60 employees,[34] continues to be pointed to as one of its most active, successful, and powerful actors in sustainable development in the country.[35] As IRDNC began to experience some successes helping communities to manage wildlife and habitat more sustainably with the backing of the power of international capital from a proliferating assemblage of donors, the notion of community-based conservation began slowly gaining traction and others began joining the movement and developing further organizations to work on similar initiatives around the country. IRDNC began to be a central actor with legitimacy afforded to it by its position as an inroad for the international community at the local level into Namibian communities. For their part, many communities welcomed the resources becoming available to them and began working to benefit however they could from proliferating

[33] Interview, Dr Ekkehard Klingelhoeffer, Head of Department—Wildlife Management and Ecotourism, University of Namibia, January 13, 2020; Interview, Michael Knight, Transboundary Leader for KAZA, WWF, January 23, 2020; Interview, Willie Boonzaaier, Programme Director, Integrated Rural Development and Nature Conservation, January 24, 2020; Interview, Brian Jones, Environment and Development Consultant (Helped draft 1996 Nature Conservation Act), Windhoek, February 20, 2020.

[34] "Integrated Rural Development and Nature Conservation - Our Team," accessed November 22, 2021.

[35] Interview, Maxi Pia Louis, Director, Namibia Association of CBNRM Support Organizations, January 14, 2020; Interview, John Hazam, Senior Technical Advisor: Namibia Nature Foundation, January 23, 2020; Interview, Manager, Palmwag Lodge, February 5 2020.

conservation efforts in the country.[36] While power imbalances that had been exacerbated by the colonial era as laid out above limited the ability of local communities to achieve certain ends, they also were being afforded increased power based on them as perceived protectors of their local wildlife populations seen as a global good. As the importance of wildlife and the global environmental movement more broadly gained traction, so too did the ability of local African populations to leverage global capital to realize sustainable development goals in their communities.

Namibia gained its independence only in 1990 after a protracted and violent fight for freedom. For the country this meant a chance at a new beginning and a new constitutional backing from which to attempt to correct the indecencies of the colonial and apartheid legacies. Independence presented development actors and the people of Namibia with a real opportunity as they would write a new constitution that could provide a legal framework from which to achieve the goals of sustainable development and community empowerment through wildlife conservation. As discussed in the previous section, prior to independence land in Namibia was extremely unfairly distributed along with the resources on that land and laws largely favoured white commercial farmers.[37]

With these injustices in mind, and partly thanks to the many lessons of history, Namibia's constitution became recognized as one of the most progressive in the world specifically for its legal requirements for environmental conservation tied to each of its major components.[38] The constitution has sections throughout outlining the need for environmental conservation which laid the legal foundation for the 1996 Nature Conservation Amendment Act that would be the formal legislation drafted to enable the CBNRM programme.[39] This legislation was written by the newly elected Southwest Africa People's Organization (SWAPO) government, but rather than being formulated entirely by the state bureaucracy the government did so in consultation with civil society, NGOs, a small army of consultants, and with input from international institutions and

[36] Interview, Uisa, Manager, Sesfontein Conservancy, February 6, 2020; Joglinde Touros, Manager, Uisbasen/Twyfelfontein Conservancy, January 30, 2020; Emil Roman, Manager, Torra Conservancy, February 2, 2020.

[37] Richardson, "Wildlife Utilization and Biodiversity Conservation in Namibia."

[38] Snively, "Environmental Laws of Namibia."

[39] Hoole, "Place-Power-Prognosis."

foreign governments.[40] While a variety of actors contributed to the process, multiple scholars point to USAID as being a key player, providing the financing required for the process as well as a great deal in the way of technical and legal input on the development of this legislation.[41] USAID worked with the Namibian government, as well as consultants and some of the NGOs which had already emerged on the ground such as IRDNC, in order to consult with communities to determine what form the law should take and how this reallocation of land and resources ought to be operationalized as well as how the restructuring of power should materialize. The new legislation also codified much of what was already in place in practice as there were a number of existing laws concerning conservation. However, existing laws consisted of a patchwork of local regulations which the Conservancy Act codified into a single set of laws that could be uniformly applied across the country to become more effective and more easily enforceable.[42] This confluence of previously individual actions coalesced into a concerted set of actions which led to this first concrete outcome for the assemblage of conservation actors.

The legislation provided a legal means for communities to voluntarily come together to agree on a delineated territory, group size, constitution, and benefits distribution plan which can be seen in Fig. 4.1. From there they could be gazetted by the MET to be officially deemed a conservancy. The MET is the main government ministry with direct oversight of CBNRM in Namibia, though it does work closely with a number of other ministries on relevant issue areas. The MET employs over 1000 staff members across the country, 900 of whom will be out in the field at any given time.[43] Conservancies were then intended to become sub-state political entities with quasi-sovereign control over wildlife within their boundaries. They were also intended to enable a democratization of resource governance so that local-level institutional capacity could emerge to bring more

[40] Lelani M. Mannetti et al., "Identifying and Categorizing Stakeholders for Protected Area Expansion around a National Park in Namibia," *Ecology and Society* 24, no. 2 (July 1, 2019): 5.

[41] Jones, "Policy Lessons from the Evolution of a Community-Based Approach to Wildlife Management, Kunene Region, Namibia," 1999; Nuulimba and Taylor, "25 Years of CBNRM in Namibia"; Boudreaux, "A New Call of the Wild."

[42] Interview, Uatirohange Tjiuoro, Ministry of Environment and Tourism: Directorate of National Parks and Wildlife, January 20, 2020.

[43] Interview, Uatirohange Tjiuoro, Ministry of Environment and Tourism: Directorate of National Parks and Wildlife, January 20, 2020.

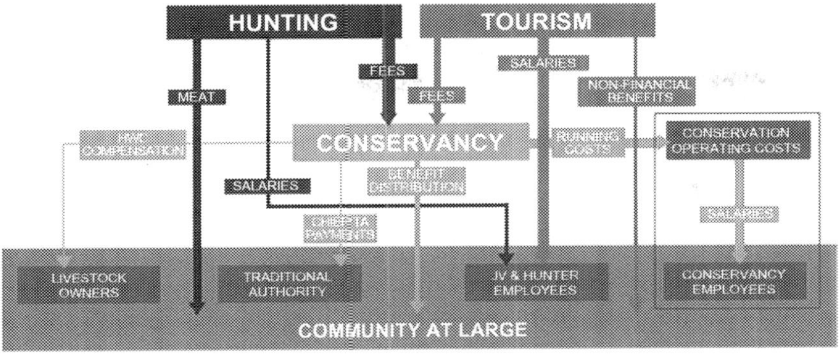

Fig. 4.1 Benefits flows in conservancies (MET/NACSO, "The State of Community Conservation in Namibia: Annual Report 2019: A Review of Communal Conservancies, Community Forests and Other CBNRM Initiatives" (Ministry of Environment and Tourism, 2021))

broad-based and equitable benefits than had traditionally been the case.[44] Tourism had long been an important industry in Namibia; however, it had always been limited to certain areas and due in large part to the legacy of Apartheid benefits often remained with white land holders, investors, and business owners.

CBNRM was intended to make the tourism industry more prolific, while also generating additional benefits and opportunities for those traditionally most marginalized in society such as women, those less well off socio-economically, under-educated people, or those lacking English proficiency. These are also goals that filtered in from the international community as domestic conservation was increasingly being realized through global partnerships and the ideologies of the international community began to materialize in development initiatives and discipline the actions of communities. One example of this was described by an MET official who said, "Namibia didn't use to care about gender issues but these became big internationally so important donor funds began being tied to gender so CBNRM has developed with a strong gender component."[45] This demonstrates the disciplinary power the international community

[44] Taylor, "Community Based Natural Resource Management in Zimbabwe."

[45] Interview, Uatirohange Tjiuoro, Ministry of Environment and Tourism: Directorate of National Parks and Wildlife, January 20, 2020.

exercised over the programme and local actors through its very genesis, as the provision of resources was either directly or indirectly tied to certain development activities and outcomes. This is where we see the complexities of the assemblage of actors involved emerge, as many scholars point to the influence of global actors as a problem, though as with the example of pushing for the importance of gender equality within the programme, there are many instances of positive global influence at the local level. However, in order to accomplish the broader goals of CBNRM a diverse assemblage of actors were empowered which resulted in a confluence of interests beyond rural marginalized Namibia populations.

Above, it was discussed that in some ways the newly formed Namibia state continued to exert sovereign power by maintaining the fortress conservation approaches implemented in the form of national parks. However, as shifts in global environmental governance as discussed in Chap. 2 increasingly called for participation and ownership, the key actors in developing CBNRM legislation in Namibia demonstrated the productive power that could be achieved by offering specific routes through which communities and citizens could become implicated in the responses to their own challenges. This was in contradiction to the previous approaches to governing that so often saw state-led approaches to conservation as well as broader issue areas. While, as is laid out above, it was an entirely voluntary process by which communities could become gazetted to become formal conservancies, there was increasing pressure as well as incentives for doing so as will be further unpacked in the next section.

Assembling Legislation for Sustainable Development

This section will further analyse the formation of CBRNM as well as the ways in which its constituent parts continue to govern and evolve following the drafting of the 1996 legislation. The Nature Conservation Amendment Act formalized the CBNRM programme and enabled its proliferation throughout the country. As mentioned, there is disagreement on which of the three pillars CBNRM was initially created to realize depending on what type of actors one asks. This results in a varying degree of emphasis being placed on each goal in various regions of the country based on the individual interpretation of actors within the policy assemblage in individual conservancies. Figure 4.2 provides an idea of some of the key NGO partners, the services they provide, and the regions in which their operations focus. One of the earliest and most useful ways

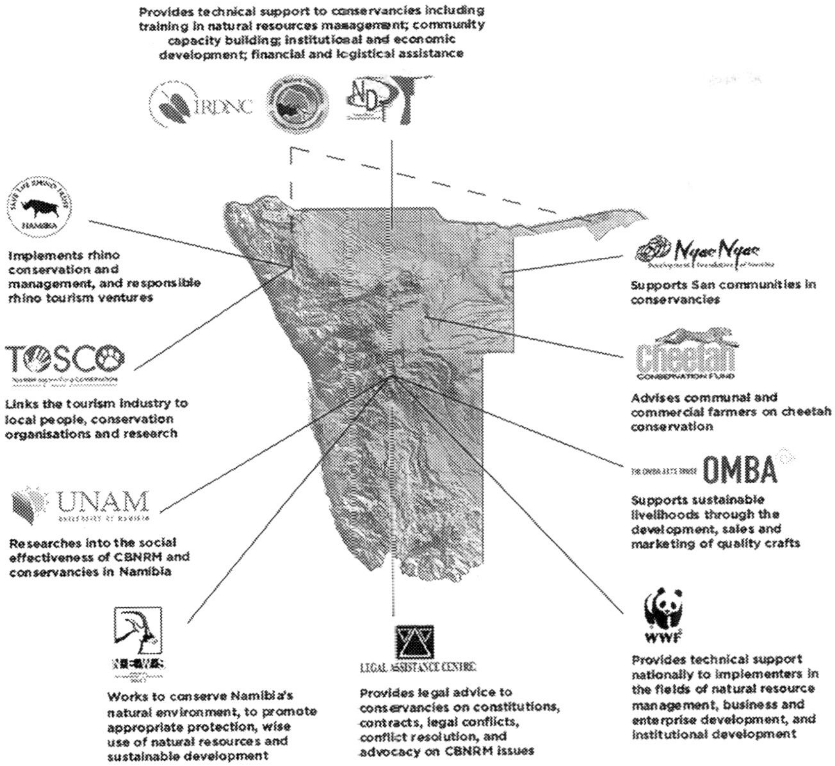

Fig. 4.2 Geographic breakdown of NGO regions (MET/NACSO, "The State of Community Conservation in Namibia: Annual Report 2019: A Review of Communal Conservancies, Community Forests and Other CBNRM Initiatives")

to outline CBNRM in Namibia as an assemblage is through the actual drafting of the Act. This legislation was drafted by SWAPO, as is all national legislation in this democratic state.[46] However, in this case the

[46] Although Namibia generally ranks relatively well in democratic and governance indicators, it has been a de facto one-party state since independence with SWAPO governing now for over three decades, which undermines a key component of most definition of democracy, namely regime change.

process was more complex than usual.[47] As the legislation was intended to empower communities, the government engaged in broad consultation with civil society in order to understand the needs and desires of rural Namibians.[48]

This consultation process was one of the first official examples of the state and its partners seeking to exercise productive power by 'educating' communities about the ways in which they ought to be implicated in wildlife conservation in order to realize their own sustainable development. Doing so would see global partners realize their goals of wildlife conservation, while also enabling the state to point to developmental results in order to garner more global standing and the resultant funding that often accompanies such results, while it was suggested communities would also be the main winners through the benefits derived from ecotourism. While this notion of consulting rural farmers and villagers is both honourable and necessary for success, one of the issues that continues to arise is a lack of education as well as both inadequate formal and informal training and experience in dealing with financial matters or management of economies beyond the household or village. Thus, as development workers and NGOs spent months and at times years consulting with communities about their needs and how best to develop CBNRM legislation, much of this time was spent discussing the logistics of the shift to CBNRM and the benefits that could be derived.[49]

Rather than the state leveraging its coercive power to enforce conservancies on rural Namibians—which through fortress conservation had demonstrated ongoing failure—the state partnered with NGOs, consultants, private tourism companies, foreign governments, and domestic environmental activists and practitioners to discipline these populations through the promise of sustainable income, development, and power for some. John Hazam explained that for some of the earliest conservancies beginning in 1994, consultation was a two-to-three-year process and in most cases led a virtual consensus within the communities as this was largely how decision making took place—through deliberation and

[47] Brian TB Jones, "Policy Lessons from the Evolution of a Community-Based Approach to Wildlife Management, Kunene Region, Namibia," *Journal of International Development: The Journal of the Development Studies Association* 11, no. 2 (1999): 295–304.

[48] Interview, Pohamba Shifeta, Minister of Environment and Tourism, January 29 2020.

[49] Jones, "The Evolution of a Community-Based Approach to Wildlife Management at Kunene, Namibia."

consensus building.[50] Despite consensus building being a key component of decision making in traditional Namibian communities, they also have traditional authorities who hold power based on generational lineages and these leaders were often inroads for consultants and NGO staff to work with in order to garner community support.[51] A wealth of research from Kalvelage et al. outlines how the rule of traditional authorities in Namibia can vary as, for example,

> the traditional authority is always specified in Zambezi [the Northeast of Namibia] whereas in Kunene, the tapestry of traditional authorities is less hierarchical... In Kunene Region, there are approximately 40 to 70 traditional authorities, some of whom compete for influence and territory. Occasionally, there are two or even more competing traditional authorities in a conservancy. This has led to the breakaway conservancies under competing chiefs in a few instances. Similar to the pre-independence period, the boundary-making process is closely entangled with the political interest of traditional authorities.[52]

Thus, we can see that neither the formation of conservancies nor their ongoing management was apolitical. Traditional authorities collaborated with NGOs in order to develop boundaries and form conservancies and have since continued these partnerships for ongoing capacity development. This is seen as a process that not only benefits local communities, but can also solidify the power and authority of traditional authorities.

To aid with the technical knowledge required to move from community consensus to gazetted conservancy, a broad array of consultants were engaged to help with the consultations as well as the actual drafting of the legislation.[53] These consultants were hired by the Namibian government but with core funding and input provided from USAID who made an initial 10-year investment in the programme, the World Bank, the WWF, as well as from a wealth of other NGOs, private donors, and many EU nations.[54] This was a time when the US was in the process of establishing

[50] Interview, John Hazam, Senior Technical Advisor: Namibia Nature Foundation, January 23, 2020.

[51] Nuulimba and Taylor, "25 Years of CBNRM in Namibia."

[52] Linus Kalvelage et al., "Territorialising Conservation: Community-Based Approaches in Kenya and Namibia," *Conservation and Society* 19, no. 4 (2021): 285.

[53] Jones, "The Evolution of Namibia's Communal Conservancies."

[54] "Funding Partners," NACSO, October 11, 2015.

its place as global hegemon, while trying to do so within the ethos of neoliberalism and moving away from direct coercive action. As a result, it leveraged its financial might to employ actors seen generally as neutral, legitimate, and in some senses depoliticized to work with partners on the ground to realize global conservation and development goals while in some ways indirectly furthering Western interests and reach. Resources were being funnelled to the Global South to realize a variety of development goals including sustainability, conservation, and good governance. Namibia was one of the benefactors as it was seen to be particularly well situated for success as a nascent democracy with a small population and large natural resource wealth. One of the key organizations to emerge from this coalescence of resources and activity was the Namibian Association of CBNRM Support Organizations (NACSO) in 1998. Today NACSO forms an umbrella organization composed of nine full members,[55] seven associate members,[56] and five individual members,[57] that are key actors in CBNRM management across Namibia. NACSO is not intended to be a central decision-making body but instead a secretariat for coordination across the many actors involved. NACSO itself and its members receive their funding from private donors, EU nations (predominantly Germany, Netherlands, Belgium, Sweden, and Norway), USAID, the UK, the UNDP, the Millennium Challenge Account (MCA), and the WWF that is their predominant funder as well as a non-voting member of NACSO.[58] This is one of the clearest ways we can see global influence taking an increasing foothold in local conservation with this global capital

[55] The full list members include: Cheetah Conservation Fund (CCF), Integrated Rural Development and Nature Conservation (IRDNC), Legal Assistance Centre (LAC), Namibia Development Trust (NDT), Namibia Nature Foundation (NNF), Nyae Nyae Development Foundation of Namibia (NNDFN), Omba Arts Trust (OAT), Save the Rhino Trust (SRT), and the University of Namibia's Multi-disciplinary Research Centre and Consultancy (MRCC-UNAM).

[56] Associate members include: Namibian Environment and Wildlife Society (NEWS), Southern Kunene Regional Association (SKRA), Tourism Supporting Conservation (TOSCO), World Wildlife Fund in Namibia (WWF in Namibia), Kavango Regional Conservancy Association (KRCA), Kunene Regional Community Conservancy Association (KCCA), and Otjozondjupa Regional Conservancy Association (ORCA).

[57] Individual members include: Anna Davis, Annie Symonds, Brian Jones, Carol Murphy, and Dhyani Berger.

[58] Interview, Maxi Pia Louis, Director, Namibia Association of CBNRM Support Organizations, January 14, 2020.

empowering NACSO as a national Namibian body that serves to flatten the previous divide between local and global actors.

This donor aid was initially allocated during a period when the international community was attempting to enact policies that could avoid centralized state action and develop novel approaches that would leverage NGOs to deliver programmes. As a result, rather than administering resources through direct channels and bilateral engagements—as would have traditionally been the case—there quickly emerged an entire network of NGOs that carved out their own geographic and issue-area spaces in Namibia.[59] This network of NGOs went to work attempting to realize development through the conservation of natural resources for Namibians and supposedly by Namibians. The NGOs worked with the Namibian government, procured funding internationally, and began to develop relationships with Namibian communities in order to gain legitimacy in the eyes of donors, all levels of government, as well as the communities themselves.

Beyond the technical knowledge and support that were needed to draft the initial legislation as outlined above, communities required ongoing support to get projects up and running. Due to persistent elevated poverty rates, many Namibians, especially in rural areas, did not possess the means to invest the time, energy, and resources in these types of conservation initiatives. High levels of inequality meant the majority of rural communities were struggling for basic subsistence during this period and could not afford to take time from their farming activities to put into something else and had no spare resources with which to invest in the development of these projects.[60] This lack of resources is one of the key contributors to power imbalances between communities and larger actors such as foreign governments, NGOs, and private companies who could offer resources as a way to demonstrate the benefits that could be gained through new ways of thinking such as wildlife conservation for ecotourism purposes.

[59] Interview, Dr Ekkehard Klingelhoeffer, Head of Department—Wildlife Management and Ecotourism, University of Namibia, January 13, 2020; Interview, Maxi Pia Louis, Director, Namibia Association of CBNRM Support Organizations, January 14, 2020; Interview, Uatirohange Tjiuoro, Ministry of Environment and Tourism: Directorate of National Parks and Wildlife, January 20, 2020; Interview, Richard Diggle, Business/CBNRM Specialist, WWF, January 24, 2020; Interview, Brian Jones, Environment and Development Consultant, February 20, 2020.

[60] Caroline Ashley et al., "Using Resource Economics for Natural Resource Management: Namibia's Experience," 1997.

CBNRM is a system of resource governance that is not as resource intensive as the building of large infrastructure projects for instance. However, it still takes a minimum amount of resources including financial and technical expertise, infrastructure, and equipment to get started. For this reason, the network of NGOs that fanned out across the country partnered with individual conservancies to provide education and assistance to help get conservancies started from the outset. Often this was a long and slow process as many within the community had little understanding of capitalistic modes of production and ordering. Thus, for many, the commoditization of wildlife was unclear and they remained sceptical, as the wildlife had always been seen predominantly as either a source of food or a source of threat.[61] The first task for the NGOs then was to engage with communities and determine what long-term impacts investments in the CBNRM programme could have.[62] In some cases, this took years, and in 1998 the first conservancies agreed and began to organize to become formally gazetted.

The consultation process was lengthy and complex. NGOs were working to achieve the goals of the global environmental movement as well as broader sustainable development goals. They were doing so through resources supplied by the international community which in this case predominantly meant the West. NGOs began to cultivate relationships with communities in order to work with them to decide how best they could benefit from CBNRM in order to harmonize local and global interests. At first, many within the communities were uninterested as they did not agree with the logic of CBNRM and ecotourism. However, as many of their neighbours grew interested, community leaders emerged that began assisting in spreading the perceived merits of this form of resource governance that was novel to Namibia.[63] In many cases these were local traditional authorities who had been entrusted with power for generations through lineage lines, while in others there were various forms of big men or community leaders who had developed respect and good standing in communities for a variety of reasons. With the proposal of CBNRM along with economic and community development through conservation, there was also the promise of new political power that could emerge through

[61] Interview, John Hazam, Senior Technical Advisor: Namibia Nature Foundation, January 23, 2020.

[62] Margaret Jacobsohn, *Life Is Like a Kudu Horn: A Memoir* (Jacana Media, 2019).

[63] Brown and Bird, "Sustainable Natural Resource Management in Namibia."

the development of conservancy councils. With these would come resources including paid jobs, conservancy offices, housing, vehicles, and more. In this way, NGOs with their own conservation aspirations procured funding from the global community with promises to realize their goals of sustainable development, while doing so through communities who themselves worked with one another in order to garner resources to realize these development initiatives. This is the way productive power functions; through bringing subjects on board in order for them to become central to their own governance by responsibilizing their actions through the capillary like processes and practices that together formulate the state and more broadly global environmental governance. Once communities began to engage with this new ethos of wildlife conservation over several years of engagement in most cases, the process required continued partnership from these various groups to realize next steps.[64]

This lengthy process took ongoing resources and support from NGOs as the legislation outlined that conservancies must have a clear boundary, group size, and formal written constitution. Conservancies became such once officially gazetted by the MET. As it was left up to the people to decide on their own membership and geographic boundaries, they collectively decided who would formulate a conservancy—though this was also impacted to a degree by NGO influence who leveraged their expertise and capital in order to influence outcomes. This is not, however, to say that they did so inherently in any sort of nefarious sense, as some and perhaps even most likely had altruistic motives. Instead, it is to begin to analyse the way existing power imbalances can lead to certain interests being realized over others. For conservancies, one could consist of a small community of 200 people, while others might span a larger geographic area and include numerous communities which could total up to several thousand conservancy members. Once formally gazetted, conservancies would become legal sub-national entities similar to a municipality, however their revenues are raised only through conservancy—predominantly ecotourism—activities. Conservancies could consist of what we might think of as a single village/township or an amalgamation of several similar such entities. NGOs worked with communities to write a constitution that detailed a resource-sharing strategy, plans for AGMs and other meetings to discuss priorities, plans, and news, as well as importantly when and how elections

[64] Ibid.

would be held to decide on a conservancy council.[65] Due in no small part to the years of Apartheid rule, these were all things community members had very little experience with. Thus, in this early stage we begin to see the way the CBNRM programme emerged in a fashion that necessitated partnerships between complex assemblages of actors in order to plan, consult, fund, operationalize, and execute the projects. While the actors involved in developing CBNRM were clear that they wanted to empower populations for their own development, it was similarly clear that these same actors saw themselves as integral for the process, and that said process ought to unfold and take shape in particular ways. In this sense, community-based initiatives demonstrate a marked improvement on previously failed cookie-cutter solutions to development, yet also remain constrained by some of the approaches that see outside actors as central to the development puzzle.

As conservancies began to be gazetted as demonstrated in Fig. 4.3 and Table 4.1, resources continued to be channelled into the country based on early signs of success and were distributed by NGOs who would assist conservancy councils in building offices and paying staff salaries as well as those of community game guards.[66] NGOs themselves were granted no specific legal right or status, but through a recognition of their knowledge and experience, they were afforded the tools, resources, and legitimacy needed to work with conservancies to realize shared goals. NGOs found those who were well-positioned within communities which tended to be those best educated or stemming from the most money, as well as those who simply demonstrated the best English language skills. As a result, those traditionally most marginalized in society tended to remain as such and their voices were heard from very little during this transition process.[67] Marginalized populations include certain minority ethnic groups in some conservancies, women, those lacking English language skills, people socio-economically disadvantaged in any number of ways, and especially intersectional people identifying with more than one of these groups.

[65] Interview, Dr Margaret Jacobson, co-founder IRDNC, co-director, trustee, mentor, Conservancy Safaris Namibia, February 17, 2020.

[66] Interview, Lara Diez, Nyae Nyae Development Foundation of Namibia, January 10, 2020; Interview, John Hazam, Senior Technical Advisor: Namibia Nature Foundation, January 23, 2020; Interview, Brian Jones, Environment and Development Consultant, February 20, 2020.

[67] Hoole, "Place-Power-Prognosis."

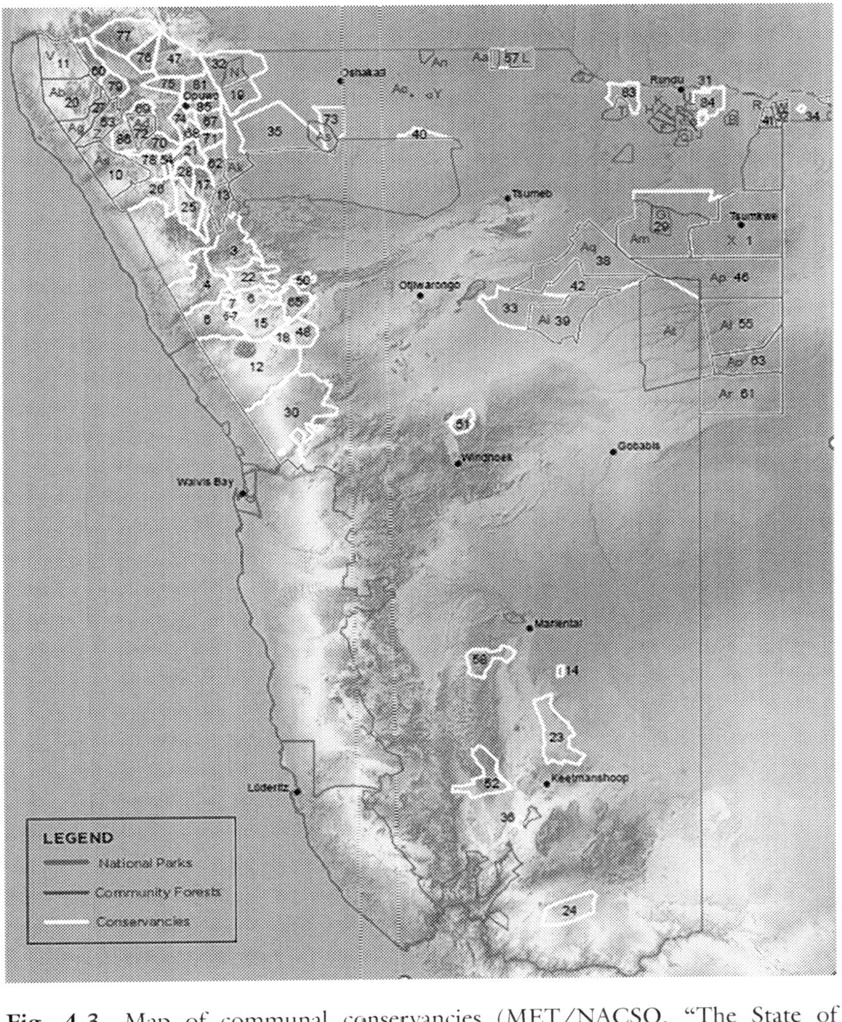

Fig. 4.3 Map of communal conservancies (MET/NACSO, "The State of Community Conservation in Namibia: Annual Report 2019: A Review of Communal Conservancies, Community Forests and Other CBNRM Initiatives")

Table 4.1 List of communal conservancies (MET/NACSO)

Map no.	*Name*	*Approx. people*	*Reg. date*	*Contact*
17	Omatendeka	2539	March 2003	081-283 7564
75	Ombazu	3207	May 2012	081-431 6825
81	Ombombo-Masitu	2962	October 2014	081-419 9982
70	Ombujokanguindi	701	February 2012	081-349 5376
63	Omuramba ua Mbirda	513	March 2011	081-298 7150
46	Ondjou	2969	October 2006	081-229 2587
69	Ongongo	881	February 2012	081-632 9117
20	Orupembe	187	July 2003	081-722 8590
62	Orupupa	1494	March 2011	081-246 8197
14	Oskop	75	February 2001	081-314 2420
54	Otjambangu	1872	March 2009	081-353 2935
78	Otjikondavirongo	3669	March 2013	081-565 4765
86	Otjikongo	210	August 2018	081-434 2270
18	Otjimboyo	323	March 2003	081-658 6055
85	Otjindjerese	2062	August 2018	081-472 3411
60	Otjitanda	575	March 2011	081-283 9550
38	Otjituuo	5921	September 2005	081-229 2587
72	Otjiu-West	831	May 2012	065-685 160
68	Otjombande	1633	February 2012	081-260 4556
61	Otjombinde	4782	March 2011	081-322 4923
71	Otuzemba	457	February 2012	081-565 2855
51	Ovitoto	4495	May 2008	081-224 4721
33	Ozonahi	11381	September 2005	081-749 1466
28	Ozondundu	395	July 2003	081-459 0310
10	Puros	1163	May 2000	081-664 2020
2	Salambala	8923	June 1998	081-824 8399
27	Sanitatas	148	July 2003	081-740 3987
26	Sesfontein	1839	July 2003	085-657 8118
34	Shamungwa	140	September 2005	081-692 0035
35	Sheya Shuushona	3542	September 2005	065-25 2088
56	Sikunga	2476	July 2009	081-252 0968
49	Sobbe	1085	October 2006	081-321 5917
15	Sorris Sorris	950	October 2001	081-397 1340
4	Torra	1330	June 1998	081-759 8470
12	Tsiseb	2645	January 2001	081-231 1371
7	Uibasen Twyfelfontein	230	December 1999	067-68 7047/8
32	Uukolonkadhi Ruacana	35958	September 2005	081-347 6455
19	Uukwaluudhi	983	March 2003	081-485 1784
5	Wuparo	1027	December 1999	081-580 6344

Once conservancies had their formal structures in place, the next set of challenges would turn towards becoming self-sustaining and eventually profitable. This is where resource management enters and specifically the commoditization of wildlife. The 1996 Nature Conservation Act outlines provisions for conservancies to engage in joint-venture agreements with private tourism companies to build lodges, camps, and conduct tourism activities in the area.[68] The logic for this was simple—the more wildlife any conservancy had, the more valuable tourism would be in their area and thus, the more they could benefit financially. In conducting interviews one of the major limitations almost every respondent pointed to was the lack of financial literacy at the community level which placed them at a difficult position for relative bargaining power. However, over time community members are increasingly engaging in training and developing skills and experience with these issues in order for them to continue to gain more ownership over their own development, specifically with these important financial issues. Nonetheless, the lack of financial literacy born in large part by the colonial legacy led to the next ongoing role played by the growing network of NGOs representing global interests which was to oversee the crafting of joint-venture agreements.[69]

The way these agreements work is that a private tourism operator—which can be either domestic or international—will approach a given conservancy with a plan for development in their area. As legal entities, conservancies are granted rights to engage in these agreements and make decisions about CBNRM activities within their borders. While in practice this right is generally respected by all levels of government, in theory state authorities maintain veto power over development decisions on the land and while communities are empowered in many ways, the state maintains the ability to exercise sovereign power should it deem necessary to do so. For joint-venture agreements, private enterprises present a proposed plan for the size of the operation along with number of beds, plans for the sourcing of food and water, transportation, as well as planned overall income.[70] Sarah Roos, manager of the Brandberg White Lady Lodge in Tsiseb Conservancy, reported, "All our decisions go through Tsiseb Conservancy council like when deciding to expand, build tree houses,

[68] "Nature_Conservation_Amendment_Act.Pdf," accessed March 18, 2020.

[69] Interview, Maxi Pia Louis, Director, Namibia Association of CBNRM Support Organizations, January 14, 2020.

[70] Interview, Manager, Palmwag Lodge, February 5 2020.

anything that could negatively impact the environment."[71] From there it is up to the conservancy to make a deal for what their share of the profits will be. This usually takes the form of either a monthly flat fee paid, a bed levy based on number of nights stayed by tourists at the lodges or campsites, or a percentage of overall profits.[72] The agreements usually have stipulations also for the percentage of local people that must be employed as well as outlining the number of years the agreement lasts before it must be renegotiated. While, as pointed to by Roos, decision making goes through conservancies, companies like the one she works for have in some cases become the predominant source of income for many communities. As a result, as seen globally with large corporations, increasingly private interests can dictate outcomes due to the importance of the jobs and GDP that come with their presence in a given community, country, or even region.[73]

These agreements are often complex, and a great deal of costs and benefits must be weighed when making decisions during the negotiation process. Ashley argues,

> A wide range of costs and benefits need to be taken into account. In particular, it is as important to address negative impacts as to maximise positive ones. And to address impacts on people's assets and existing activities, not just direct contributions to household income and security.[74]

She develops research based on a livelihoods approach that when analysing CBNRM benefits broadly, and in conservancies engaging in joint-venture agreements specifically, it is important to look at much more than simply 'jobs and cash.' Her research suggests it is integral to also look at a much broader range of impacts that are key to inclusive and sustainable livelihoods for local populations.

[71] Interview, Sarah Roos, Manager, Brandberg White Lady Lodge, January 28, 2020.

[72] Interview, Uatirohange Tjiuoro, Ministry of Environment and Tourism: Directorate of National Parks and Wildlife, January 20, 2020; Interview, Manager, Okahirongo Elephant Lodge, February 8, 2020; Interview, Christine, Manager, Anabeb Conservancy, February 7, 2020.

[73] Philip G. Cerny, "Paradoxes of the Competition State: The Dynamics of Political Globalization," *Government and Opposition* 32, no. 2 (1997): 251–74.

[74] Caroline Ashley, "The Impacts of Tourism on Rural Livelihoods: Namibia's Experience" (Overseas Development Institute London, UK, February 2000), 6.

As the CBNRM programme is built around the ethos of conservation, and communities are intended to realize development based on the responsible management of local wildlife and habitat, it becomes essential as well to have strict limits written into these agreements regarding what tourism companies can build, how they can do so, how they will source materials as well as when and how they can expand future operations.[75] All of this adds up to very long and complex agreements that many Namibian community leaders possess inadequate knowledge and skills to fairly engage in with private companies with years of experience. This is one of the issues that devolution can create. States generally have ministries, a diplomatic corps, and a specialized bureaucracy to staff these agencies and negotiate trade deals, as well as agreements with private firms—whether small-scale enterprises or mega transnational conglomerates—rural communities will not have the same ability to bargain, nor the human resources needed to effectively do so. This exacerbates existing power imbalances and while the legislation is intended to place full control in the hands of conservancies represented by their elected councils, they can only accomplish their ends, by way of partnership with private interests.[76]

Often what this results in is international capital being able to negotiate deals that are favourable to its backers, and less so to local communities—even when legislation is intended to favour their empowerment. While there are power imbalances between communities and private tourism organizations, as well as conservancies and NGOs, conservancies remain at the centre of CBNRM and are legally provided the mandate over wildlife. As a result, these groups—and indirectly the international community—require local communities to become effective conservancies in order to partner with them for the protection of wildlife. Some of the indicators used for measuring this are displayed in Table 4.2. In this sense, CBNRM policy provides a legal framework for the empowerment of communities, yet in order for them to do so they must paradoxically enlist the support of external actors to realize their individual goals, in what at times can be synergistic—though of course it is not always so. This demonstrates not only how the policy assemblage develops, but also why it is necessary

[75] Interview, Joglinde, Manager, Uisbasen/Twyfelfontein Conservancy, January 30, 2020; Interview, Emil, Manager, Torra Conservancy, February 2 2020; Interview, Dr Margaret Jacobson, co-founder IRDNC, co-director, trustee, mentor, Conservancy Safaris Namibia, February 17, 2020.

[76] Mercer, "Performing Partnership."

Table 4.2 Governance indicators for conservancies (MET/NACSO, "The State of Community Conservation in Namibia: Annual Report 2019: A Review of Communal Conservancies, Community Forests and Other CBNRM Initiatives")

Cat. or sub-cat.	*Category (N = number reporting in 2019)*	*Number*	*Percentage of category*
1	Registered conservancies (incl. Kyaramacan assoc.) (N = 67)	87	100%
2	Conservancies generating returns (N = 87)	65	75%
2a	– Covering operational costs from own income (N = 54)	30	56%
2b	– Distributing cash or in-kind benefits to members, or investing in community projects (N = 54)	43	80%
3	Conservancy management committee members (N = 82)	975	100%
3a	– Female management committee members (N = 82)	320	33%
3b	– Female chairpersons (N = 82)	11	13%
3c	– Female treasurers/financial managers (N = 82)	35	43%
4	Conservancy staff members (N = 82)	964	100%
4a	– Female staff members (N = 82)	232	24%
5	Conservancies management plans (N = 82)	57	70%
5a	– Sustainable business and financial plans (N = 82)	27	33%
6	Conservancy AGMs held (N = 82)	67	82%
6a	– Financial reports presented at AGM (N = 82)	53	65%
6b	– Financial reports approved at AGM (N = 82)	50	61%
6c	– Budgets approved at AGM (N = 82)	49	60%

in many ways. It also further elucidates the many forces that are simultaneously competing and working together to realize various goals through these supposedly community-based projects.

Assembling Conservancies

As discussed thus far in this chapter, the Nature Conservation Amendment Act was developed with input from a wide assemblage of actors in order to build from the best of neighbouring examples and also to learn from their mistakes.[77] The most important of these lessons was to realize devolution

[77] Brian TB Jones, Community-Based Natural Resource Management in Botswana and Namibia: An Inventory and Preliminary Analysis of Progress (IIED, 1999).

to the lowest level possible which did not occur in neighbouring countries and which scholars from both the utilitarians and sceptics point to as a reason for their relative lack of success.[78] Therefore, Namibia's version sought to empower communities to the maximum extent. As such, the legislation lays out stringent requirements conservancies must meet before they can be gazetted. The legislation states that

> any group of persons residing on communal land and which desires to have the area which they inhabit, or any part thereof, to be declared a conservancy, shall apply therefor to the Minister [of Environment and Tourism] in the prescribed manner, and such application shall be accompanied by: (a) a list of the names of the persons who are members of a committee established for the purpose of being recognized by the minister under subsection (2)(ii) as the conservancy committee for the conservancy applied for; (b) the constitution of such committee; (c) a statement setting out the boundaries of the geographic area in respect of which the application is made; and (d) such other documents or information as the Minister may require.[79]

As part of this they must meet the requirements discussed in the previous section and an important part of this is the holding of AGMs and regular elections. As they are responsible for writing their own constitutions, these processes and the make up of conservancy leadership vary from conservancy to conservancy. However, generally conservancy councils consist of a manager who will be a full-time staff member employed by the conservancy and tasked with managing the day-to-day activities, representing the conservancy, and managing the team. In most cases conservancy staff will also include a chairperson, treasurer, office manager, head of game guards/wildlife, and other positions depending on level of revenue and need.

In Eric's conservancy, Tsiseb as one example, the conservancy council consists of nine members, none of whom outside the manager receives a salary and instead they receive N$250 per month as per diem assistance and a small travel allowance. Tsiseb, like many others, also has regulations for regional representation. The conservancy is made up of four regions and of the nine members; there are requirements for representation from each region.[80] Conservancies consist of between 200 and as many as 5000 members. Each conservancy however consists of a larger population than its voting members. Who constitutes a member and how membership can

[78] Boudreaux, "A New Call of the Wild"; Schnegg, "Lost in Translation."

[79] MET, "Nature Conservation Amendment Act," 3.

[80] Interview, Eric Xaneb, Chairman, Tsiseb Conservancy, January 27, 2020.

be attained vary based on each conservancy constitution, but generally the requirements are that members are at least 18 years of age and have lived in the conservancy for a minimum of 5 years. These requirements are put in place to avoid people switching conservancies when one is doing better than the next and might be providing more in the way of benefits for its people. The percentage of the population that are voting members ranges from as low as 40 per cent to as high as 80 per cent, with the average falling somewhere in the middle. Full conservancy members are afforded the right to vote, as well as access to any benefits derived through the conservancy. Non-members are excluded from both of these rights and, thus, membership is generally highly sought after which has also led to growing populations in the majority of conservancies.[81]

Elections vary among conservancies, occurring every three to five years and with most constitutions limiting conservancy leaders to a maximum of two or three consecutive terms in office, though in some there are no term limits. AGMs are an important and necessary component of each constitution. At AGMs elections occur, but also annually the conservancy council must present a benefits distribution plan to the people who must vote on whether to accept it or not. In this sense it is similar to a government's annual budget process in which the council lays out their financial situation regarding their bank account, previous year's income and spending, as well as the next year's expected revenue and spending plans. This is intended to provide conservancy members with a clear idea of governance priorities and if a majority votes against the year's resource-sharing plan, the council must consult with them in order to develop a new plan which they must subsequently present and hold another vote to pass.[82] AGMs are also where plans for new joint-venture agreements or modifications to existing ones will be presented and again a majority of conservancy members present must vote in support of any changes in order for them to be approved. This democratic process was key for many respondents suggesting they felt a sense of ownership over their own conservation and development. The Anabeb Conservancy manager asserted: "No one tells us

[81] Julie A. Silva and Alfons Mosimane, "'How Could I Live Here and Not Be a Member?': Economic Versus Social Drivers of Participation in Namibian Conservation Programs," *Human Ecology* 42, no. 2 (2014): 183–97.

[82] Interview, Emil Roman, Manager, Torra Conservancy, February 2, 2020.

what to do. Neither NGOs nor government. They just come to hear and take notes on what conservancies say they're going to do."[83]

Although the degree to which conservancy members feel their voices are heard varies both within and between conservancies, this development of democratic processes and institutions is a key component of CBNRM governance and in seeing communities effectively develop empowerment. It is also however a key example of the productive power of the state enabling communities to manage their local wildlife, but only if they first meet a stringent set of requirements including democratic processes and the ordering and structuring of local populations through publicly recorded memberships. In Namibia with such a small population spread out across a vast territory, these are functions the state has traditionally not been able to effectively enforce. CBNRM, however, enables the state to work with global partners to incentivize rural communities to in a sense project the state's power from a distance and to perform state-like functions sans state resources.

This shift in local, national, and global power relations shifts responsibilities for governance to communities which can produce benefits locally but also maintains required solutions to challenges and associated costs locally as well. While both cash and in-kind benefits have decreased significantly for many, conservancies have evolved and continue to invest in important ways in communities and their people. The MET has sought to push for this type of investment as Pohamba Penomwenyo Shifeta, the current minister, implemented a new mandate in 2018 that at least 50 per cent of conservancy revenues must be reinvested into the community. This mandate has had the effect of putting a cap on what conservancies can spend on overhead and designating a higher percentage of revenues directed to community benefits for conservancy members.[84] While this has been controversial, as many conservancies with low incomes have suggested they cannot afford to maintain staff and keep the lights on at these levels, for others, it has pushed them to increase investments in important ways. A focus group in Sesfontein Conservancy, 150 km Northwest of Torra Conservancy, had conservancy members repeatedly pointing to investments that include improvements in freshwater, power, food, ambulances, funeral costs, medical care, primary education, scholarships, and

[83] Interview, Christine, Manager, Anabeb Conservancy, February 7, 2020.

[84] Interview, Uatirohange Tjiuoro, Ministry of Environment and Tourism: Directorate of National Parks and Wildlife, January 20, 2020.

more.[85] One MET official suggested this was one of the problems that took time to shift attitudes, as it is easy for someone to link cash being handed to them to a positive outlook towards the conservancy.[86] However, when one wakes up every day and is able to turn on the light and has sufficient drinking water, or when they are injured or become ill and need to be taken several hundred kilometres to the nearest hospital and an ambulance is available to do so, these benefits are much more easily taken for granted. Conservancies also work closely together in many cases pooling resources in order to make investments. Torra, Anabeb, and Sesfontein are known as the 'Big Three' and together purchased an ambulance which was pointed to with pride repeatedly during interviews. This included one man in a focus group who proudly proclaimed

> The Big Three split on it and surprised the Ministry of Health with it. It can take people from three conservancies to the health centres, or to Opuwo [the closest city] if they need the hospital instead of an ambulance having to drive 3 h each way to pick sick people up and take them back. It is saving lives and something we are very proud of.[87]

Thus, conservancies have in many ways taken a positive step towards investing in these important services for long-term success and development—as seen in Fig. 4.4—despite the fact that these benefits may not always be associated with the programme. Importantly, this is also an example of conservancies working with a number of partners to benefit economically from wildlife conservation—highly sought globally—and from these benefits providing local community members with the types of services normally associated with the state.

When people come together into some form of political unit, they tie their futures and outcomes to one another in important ways and this also means a degree of collective decision making. While many favour longer-term investments, there will always be some who will prefer the immediate reward of cash in hand. Several interview respondents in fact suggested that this would be a useful way to quell some of the discontent regarding allegations of corruption by presenting the incomes of conservancies in the way of actual cash on the table at AGMs so that people can see what is

[85] Focus group 4, Anabeb Conservancy, February 7, 2020.

[86] Interview, Uatirohange Tjiuoro, Ministry of Environment and Tourism: Directorate of National Parks and Wildlife, January 20, 2020.

[87] Focus group 3, Anabeb Conservancy, February 7, 2020.

Fig. 4.4 Community conservation at a glance (MET/NACSO, "The State of Community Conservation in Namibia: Annual Report 2019: A Review of Communal Conservancies, Community Forests and Other CBNRM Initiatives")

theirs. In an interview, Brian Jones suggested that by physically placing conservancy incomes in the form of cash on a table at AGMs, it could from there be divided up evenly so that everyone gets a chance to hold it in their hand, and feel what is rightfully theirs, prior to then paying into the types of overhead costs that exist as well as investments that conservancies might collectively decide on.[88] While this poses potential to remedy certain issues, it is not clear that this would be an efficient form of

[88] Interview, Brian Jones, Environment and Development Consultant (Helped draft 1996 Nature Conservation Act), Windhoek, February 20, 2020.

governance and would be fraught with challenges of its own. On the other hand, Anabeb Conservancy, 12 km east of Sesfontein, is one conservancy that has been distributing very little in the way of cash remittances, but its members reported very positively on the conservancy and the many non-cash benefits they enjoy. These were seen largely in the form of local-level investments, the type which many would agree amount to sustainable development initiatives. Investments include improvements to infrastructure for procuring freshwater, medical services, funeral payments, and old age subsidies among many others.

These types of difficult resource allocation and governance decisions are at the heart of conservancy politics in Namibia and are largely determined by the systems discussed above. Conservancy councils are democratically elected by conservancy members on a regular basis. Those who wish to run for election will outline their priorities and present a platform of sorts and people vote in part based on that, though also largely on popularity and relations with the individuals. Thus, if a council's spending priorities align with the interests of members, in theory they will stand a good chance of re-election. Those who are seen to be mismanaging conservancy funds can be voted out. A police officer near Bersig reported that

> there have been political issues with the conservancy like anywhere else, but they are not that bad, and we were able to vote for a new council in December [2019] which makes us at least feel like our voices are heard. If we cannot vote for a good council, at some point it is a reflection on us, not a fault with the programme.[89]

There are a plethora of variables that determine what one conservancy chooses to spend its income on versus another, however, at the heart of decision making, is a democratic process that would be recognizable to many.

It is this type of debate that has been at the heart of CBNRM politics for years and it is one that has been taking time to shift opinions. It was also in an effort to ensure that broad-based support for the programme continued that the 50 per cent community benefit policy was implemented by the Minister of Environment and Tourism, based also on input from a number of global partners. While there has been opposition to this shift, there is evidence that the push to invest more tourism incomes back into

[89] Interview, Police officer, Bersig, Torra Conservancy, February 4, 2020.

communities has contributed to the types of demonstrable benefits outlined above.[90] These are many of the same questions that NGOs as well as charitable and non-profit organizations struggle with more broadly, as donors want reassurance their funds are going directly to helping people or a given cause, yet there is always a level of overhead needed in order to pay staff salaries, rent office space, as well as to fund travel, meetings, per diems, and many other (less sexy) expenses that come with development initiatives.

For conservancies specifically, overhead encompasses maintaining office space which includes hydro, internet, and office supplies, staff salaries for the conservancy council as well as game guards, and in some of the wealthier conservancies vehicles to provide important transportation for conservancy activities through the vast and rugged terrain.[91] These are the type of struggles which continue with many development projects, international agencies, NGOs, and even with private firms across the world. Paradoxically it seems that it has taken the existential challenges posed by climate change to convince broad sections of conservancy members to begin to shift to seeing investment in more long-term projects as preferable to shorter-term remittances. As NGOs and their partners leverage their resources, education, and experience in order to exercise productive power to try to convince populations that these are beneficial and in their own interests, they are able to point to the deleterious impacts of climate change as a necessary reason for doing so. Dr Chris Brown, CEO of the Namibian Chamber of Environment and one of the architects of CBNRM in Namibia, openly stated at a workshop with the NACSO partners that "climate change and the related ongoing drought has served as a much needed learning experience and kick in the ass to us and to the conservancies."[92] He suggests that efforts that have been underway to fight climate change have been linked to further social uplift specifically with regard to health, electricity, and water and that these efforts must be redoubled to ensure the programme continues to become more inclusive as they seek to make it more sustainable. As elaborated in research from the sceptics, while there was growing disillusionment towards the programme since 2014, as the drought has continued it seems for some there

[90] Interview, Pohamba Shifeta, Minister of Environment and Tourism, January 29 2020.

[91] Interview, Albert Kroses, Head of Environmental Services, #Khoadi !Hoas Conservancy, February 10, 2020.

[92] Interview, Dr Chris Brown, CEO Namibia Chamber of Environment, January 24, 2020.

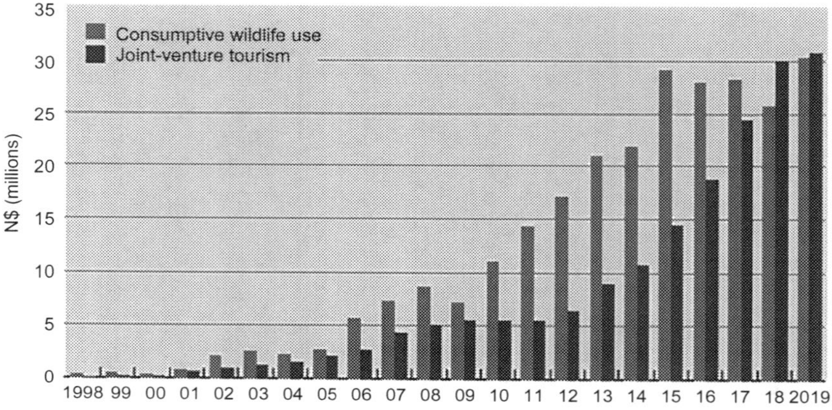

Fig. 4.5 Cash revenues from tourism operations (MET/NACSO, "The State of Community Conservation in Namibia: Annual Report 2019: A Review of Communal Conservancies, Community Forests and Other CBNRM Initiatives")

has been a reversal in the trend. Rather than blaming CBNRM for no longer providing the benefits it once did, one game officer reported:

> CBNRM is good. I am employed and I can see it [CBNRM] bringing jobs and giving us a say. People vote, we are not without our troubles but with everything there are challenges and things are in some ways getting better… The drought is very challenging but without CBNRM it would wipe us off the map. Instead with CBNRM we are coming onto the map.[93]

This game officer felt that as tourism increases as seen in Figs. 4.5 and 4.6, income was being provided in places where farming was virtually no longer possible. Moreover, he stated that even where we were standing could be found on Google Maps which was previously not the case. More and more this is becoming the norm and he thinks this is important, so that the world knows they are there.

These sentiments were echoed by a rhino game guard from Torra Conservancy employed by Save the Rhino Trust, one of NACSO's key affiliates, who said:

[93] Interview, Game officer, Torra Conservancy, February 2, 2020.

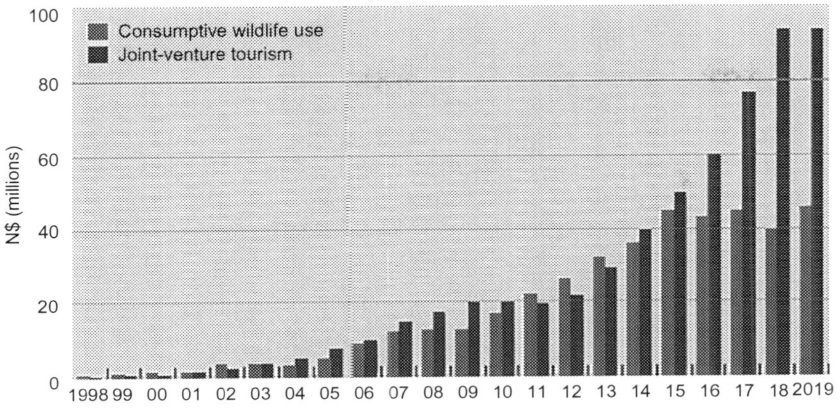

Fig. 4.6 Growth in non-consumptive tourism (MET/NACSO)

> The last council was not too bad but there were issues. Because of the drought we had to stop hunting and trophy hunting was our main source of income, so people stopped getting meat and direct cash so then they blamed the council for corruption. But the program has helped with medical care, police funding, and giving the people a place to speak their voices and vent frustration. This is probably a good healthy release valve in tough times of drought.[94]

Again, this sentiment that CBNRM was continuing to bring benefits during the drought was echoed by several voices in a focus group in Anabeb Conservancy. One young woman replied, "Health care is better. We used to get food which we can't anymore because of the drought, but at least health care is better." This quote demonstrates the way at least some conservancy members have begun to internalize these challenges in a sense, and rather than blaming others, or looking to the state, or donors for help they have been responsibilized to feel empowerment and ownership of their developmental challenges as well as the achievements some of which are outlined in Fig. 5.1. The person next to her then said, "Yes and one of the important things we get is conservation. Before no one cared about it and the animals were disappearing and their habitat being destroyed. Now people take pride in protecting them and we all do it

[94] Interview, Rhino game guard, Torra Conservancy, February 4, 2020.

together as a community." After a general agreement from the group, a middle-aged man added, "Conservation is ours and we make things happen, others do not tell us what to do. Tourism has been huge for us. With the drought tourism is everything and we get incomes from the lodges and our wives can make crafts and sell it to tourists. We very much like tourists."[95] These sentiments were further echoed by a rural respondent from Torra who reported:

> We receive benefits like water, medical treatment, education for kids. We like the program, and it is helping the community grow and get better. There are ups and downs and times are tough now but will improve. Our community are communal farmers but also sell crafts and goods out on the roads to tourists so women and children can benefit financially directly from tourism more than anything.[96]

These findings build on prior research by Ashley and LaFranchi that found

> Grass-sellers, craftmakers, fisherfolk, casual labourers and others sell their products or labour to tourists and tourism enterprises (and to neighbours earning wages from tourism jobs). They each earn much smaller amounts from these casual sales than is earned by those in waged employment (tens or hundreds of Namibian dollars each per year). But these additional earning opportunities are likely to benefit a higher percentage of local households than the full-time jobs and are most important for the poorer people who have few other options for earning cash.[97]

While much of this data points to a significant sentiment of empowerment in the community, the next section will wade through some of the tensions that continue to result in ongoing challenges (Fig. 4.7).

[95] Interview, Focus Group 1, Anabeb Conservancy, February 7, 2020.

[96] Interview, Village respondent 4, Torra Conservancy, February 1, 2020.

[97] C. LaFranchi and C. Ashley, "Livelihood Strategies of Rural Households in Caprivi: Implications for Conservancies and Natural Resource Management," Working Paper (Directorate of Environmental Affairs Ministry of Environment and Tourism, Windhoek, Namibië, 1997), 20.

Fig. 4.7 Benefits derived from CBNRM (MET/NACSO, "The State of Community Conservation in Namibia: Annual Report 2019: A Review of Communal Conservancies, Community Forests and Other CBNRM Initiatives")

Assembled Governance and Charges of Corruption

Despite a widespread sense of satisfaction with the programme there remain a number of concerns with local-level governance. This section turns to discuss allegations of corruption at the conservancy level, which contributes to addressing my second research question regarding the degree to which communities benefit as a result of CBNRM. Analysing allegations of corruption also further elucidates the layered nature of this complex policy governance assemblage. As with any system of governing CBNRM is far from perfect. Compounding challenges inherent to governing are that it is a relatively nascent and ever-evolving system of governance, with new conservancies continuing to be gazetted regularly, the lack of education and experience, and the layered approach to governance creates an inherent potential for mismanagement. While going local is suggested by some to circumvent the corruption that has been experienced by the state with resource management across much of Southern Africa,[98] ample research has made it clear that devolution in itself is far from a panacea.[99] Dr Ekkehard Klingelhoeffer reports that with CBNRM "there is corruption and mismanagement everywhere. You try to turn these projects over to the communities, but they are not educated in financial matters and business management."[100] There is however a difference between mismanagement and corruption, and while it seems that both are present within Namibia's CBNRM programme, accusations of both have also increased with the onset of the drought. John Hazam suggests that mismanagement often occurs simply as a result of poor bookkeeping due to a lack of financial literacy which is why NGOs serve an important role in providing ongoing technical assistance. Transparency International defines corruption as the misuse of power for private gain and Hazam sees this as separate from mismanagement, but also a present concern.[101] Corruption, he argues, consists of actual theft of conservancy funds and the use of public resources for private gain and he suggests this has also been occurring and continues to go unpunished.[102]

Allegations of corruption within certain well-known conservancies are prevalent and were openly discussed by many respondents building on similar reports of elite capture within the literature.[103] Torra Conservancy is widely known as one of the country's greatest success stories, but is also

[98] Elinor Ostrom, "Coping with the Tragedy of the Commons," 1999; Schumacher, *Small Is Beautiful.*

[99] See for example: Duffy, "The Potential and Pitfalls of Global Environmental Governance"; Brockington and Igoe, "Eviction for Conservation"; Agrawal and Gibson, "Enchantment and Disenchantment."

[100] Interview, Dr Ekkehard Klingelhoeffer, Head of Department—Wildlife Management and Ecotourism, University of Namibia, January 13, 2020.

[101] Transparency International, "Transparency International Report (2006) Africa," 2006.

[102] Interview, John Hazam, Senior Technical Advisor: Namibia Nature Foundation, January 23, 2020.

[103] Schnegg, Bollig, and Linke, "Moral Equality and Success of Common-Pool Water Governance in Namibia"; Boudreaux, "A New Call of the Wild"; Hoole, "Place-Power-Prognosis."

quietly referred to by some as CBNRM's problem child,[104] as well as it being suggested that it struggles with governance.[105] Torra is one of the longest running and highest earning conservancies in Namibia. While at one time the conservancy was providing a great deal for its members in the way of services and N$550 per year in cash payments to each member, the provision of services has slowed, as had the payments over a number of years, and by 2020 they had halted altogether. This has been paralleled by a direct violation of their constitution as AGMs had not been held since 2018. The first village respondent I interviewed in Bersig, the main village in Torra, said, "I am a conservancy member, I go, I vote and I like it [the programme] but don't like corruption. Some people just talk and talk and don't do any of the stuff that needs fixing."[106] On this subject, Emil Roman, the Conservancy Manager, lamented, "Money is war. You work very hard to get it but when it comes, it's war. Now you are eating the money of the community. No one trusts anyone else; you only trust your money in the bank and even then, you don't like the interest rate and move it elsewhere." He continued by saying that "we make monthly budgets and members must approve annual budgets. Whenever we overspend budget lines we ask members, but I won't lie we are not doing this right now and it is something we must improve."[107]

Emil attributes blame for this lack of consultation and the constitutional violations by ceasing to hold AGMs on climate change and the rest of the conservancy council. He says that as a result of the ongoing drought which has been caused by climate change, conservancy incomes have dwindled, while the membership has increased and he says the council wanted to pause AGMs as they knew the people would be unhappy with the lack of cash benefits and thus wanted to wait until they had funds to distribute once again.[108] One interview respondent from just outside of Bersig suggested the council "cancelled the meeting because of the drought so now it's hard for our voices to get heard. In one sense it is good because we voted for them, but now if they won't even hold

[104] Interview, Marina Tavolaro, PhD candidate in Biological Sciences at University of Cape Town, January 15, 2020.

[105] Interview, Uatirohange Tjiuoro, Ministry of Environment and Tourism: Directorate of National Parks and Wildlife, January 20, 2020.

[106] Interview, Village respondent 4, Torra Conservancy, February 1, 2020.

[107] Interview, Emil Roman, Manager, Torra Conservancy, February 2, 2020.

[108] Interview, Emil Roman, Manager, Torra Conservancy, February 2, 2020.

meetings because of drought issues what is the point?"[109] This sentiment was put more bluntly by another interviewee while also accepting some of the onus as voters stating "if snake bites you once, it happens. If he bites you again, it will be much deeper and more serious. Everyone knows who they are, so they have to be got rid of. But we know they are just one snake and that doesn't mean the program is the one biting people."[110] While members of Torra Conservancy seemed concerned about the possible embezzlement of funds and these sentiments were echoed by many in Windhoek, interestingly enough some members of local conservancies still view Torra as the goal based on its history of high incomes and broad member benefits.

Prior to arriving in Torra I met a tour guide from Uisbasen/ Twyfelfontein Conservancy who asked to be referred to only as Barnabas, who reported

> I like the conservancy programme and I am an official member and I generally attend meetings and vote. But all our conservancy committee does is just talks talks talks, ya, ya, ya, ya and that's all they do without benefitting anyone except for their families and friends. But most people don't benefit because they [the council] don't make good decisions for everyone. I hope this will change but I am not sure why we have not found a solution or voted for a better leader. Torra is very well run and does things really well, everyone benefits. I hope we can be more like Torra.[111]

One of the issues with these conflicting accounts is that the accusations of corruption remain largely hearsay and there have not been adequate official investigations into the allegations. Critics suggest this is because CBNRM has become a global calling card of sorts for Namibia and with being deemed a global leader in this form of resource governance comes the pressure for it to continue to appear outwardly as successful as possible.[112] In this sense it is important for the Namibian state for the programme to appear outwardly successful to continue to garner global

[109] Interview, Village respondent 5, Torra Conservancy, February 1, 2020.

[110] Interview, Village Respondent 6, Torra Conservancy, February 2, 2020.

[111] Interview, Barnabas, tour guide, Petrified Forest, Uisbasen/Twyfelfontein, January 31, 2020.

[112] Interview, John Hazam, Senior Technical Advisor: Namibia Nature Foundation, January 23, 2020; Interview, Marina Tavolaro, PhD candidate in Biological Sciences at University of Cape Town, January 15, 2020.

resources, support, and continue growing its tourism industry, while it is similarly seen as integral for conservancies to be viewed as successful and well run for the same reasons. Hazam reports that this is evident as there are Standard Operating Procedures that he personally worked on which are built into the CBNRM legislation that conservancies must follow. If they fail to do so, it is within the power of the MET to de-gazette them. However, he suggests this is an important tool that has never been used nor even threatened and as a result corruption is allowed to continue. Due to the nature of conservancies which are supposed to be quasi-sovereign entities, the MET is hesitant to step in regarding issues of governance which it appears in some cases has enabled traditional gatekeepers to maintain or increase their position as local elites and slowly seize control of power.[113]

This also demonstrates the limits of concepts such as sovereignty as well as how a policy assemblage approach enables us to look beyond these sorts of binaries and reification to instead analyse the complexities of global environmental governance rather than seeing conservancies as either sovereign entities or not. Richard Diggle echoed Hazam's sentiments suggesting

> There needs to be a series of checks and balances and steps to get reported to the Ministry [of Environment and Tourism] but it's always warnings and long timelines to avoid real punishment. There needs to be strict legislation with enforcement mechanisms that lay out if you do or do not do this or that you will get X fine and Y imprisonment. The Ministry has it for poachers but not for community issues. Individuals have been held accountable but there has been no punishment so then others do it. The Ministry just asks for more oversight over the whole programme which creates a whole blanket problem for all conservancies without providing the capacities to fulfil them which just drains resources. CBNRM is a set of rights that needs an effective set of responsibilities to go with them.[114]

This sentiment was shared by not only NGO officials but even some conservancy managers themselves who pointed to a lack of effective oversight that allowed for corruption by some to result in negative impacts for them all. As was discussed in Chap. 2, while disciplinary and productive power are intended to alleviate the state from governing every move of its

[113] Bollig, "Towards an Arid Eden?," 784.

[114] Interview, Richard Diggle, Business/CBNRM Specialist, WWF, January 24, 2020.

subjects, it does so partly with the threat of coercive power as a persistent possibility. Here we see the tensions these power relations can pose to competing actors which further elucidates some of the complexities of these layered governance arrangements as well as the tensions and contradictions that can result.

Conservancy councils/managers are more or less unanimous in wanting a more complete devolution of control over management and resources and land with less oversight and more autonomy. However, many of these same leaders point to allegations of corruption in neighbouring conservancies and call for stricter regulations to prevent these instances from happening which places the entire CBNRN programme in a negative light. Eric Xaneb bluntly stated:

> I could easily just take all the money and leave. Nothing is stopping me. In many cases people who have never comprehended N$100,000 or in some cases much more, and all of the sudden they are in charge of managing it effectively. This is a problem, and the government must strengthen laws and provide additional support and training.[115]

The Minister of Environment and Tourism spoke openly of local corruption and conservancy governance issues as well as the need to mitigate these. While he suggests they have been working hard to implement solutions, more work remains to be done.[116] While not voiding the state of blame, devolution also enables the MET to distance itself from the allegations and serve as an apparent neutral party seeking only to help. This is further reflected by key literature suggesting that CBNRM governance in Namibia takes the form of all carrot, with no stick.[117] As core funding has dried up, and an increasing number of NGOs and conservancies vie for an ever-shrinking donor pie, there are fears that damage to any one component of the country's broader programme will reflect negatively on all involved. Beyond concerns of the CBNRM programme being viewed in a negative light, if there is mismanagement of conservancy funds—whether due to illicit intent or not—it is community members who are most harmed and potential benefits they can receive through the programme are minimized. This is particularly problematic for marginalized popula-

[115] Interview, Eric Xaneb, Chairman, Tsiseb Conservancy, January 27, 2020.
[116] Interview, Pohamba Shifeta, Minister of Environment and Tourism, January 29 2020.
[117] Boudreaux, "A New Call of the Wild."

tions who often rely on these benefits for their very survival. However, it is also deeply concerning as devolution enables communities to benefit from resources, but also places responsibility for the costs incurred for doing so on them. The result being that in some cases communities will be incurring the costs of conservation without enjoying the subsequent rewards.

One of the ways in which CBNRM was intended to prevent corruption and mismanagement was through the decentralization of governance by having conservancies be small so that people would be connected on some level of personal relationship. While this can and has been demonstrated to have positive effects, it can also be problematic, specifically when it comes to the reporting of crime or corruption. As many conservancy leaders have been in power for an increasingly long time and were in many cases village leaders prior to the programme, some suggest they are extremely well connected within their communities, and often dole out resources within their own localized neopatrimonial webs.[118] This serves to keep those close to them happy which, mixed with the ties those within communities or families feel towards one another, may be preventing reporting what to some is obvious mismanagement at best, and blatant corruption at worst.[119]

While there is sufficient evidence to suggest that some level of mismanagement has been occurring in Torra Conservancy as well as suggestions of mismanagement to a lesser degree in other conservancies, these same interview respondents suggested that these issues are really in one specific case or only in a small percentage of conservancies, and that it is not a systemic issue.[120] Like with any system of governance anywhere in the world

> there are good leaders and bad, just like with anything else so management can wax and wane, corruption can occur, but better management can also happen and is more prevalent... There are instances of mismanagement, but the MET works to help communities remedy this themselves and we [the MET] step in where we must.[121]

[118] Child and Barnes, "The Conceptual Evolution and Practice of Community-Based Natural Resource Management in Southern Africa," 292.

[119] Schnegg, "Institutional Multiplexity."

[120] Interview, Marina Tavolaro, PhD candidate in Biological Sciences at University of Cape Town, January 15, 2020.

[121] Interview, Uatirohange Tjiuoro, Ministry of Environment and Tourism: Directorate of National Parks and Wildlife, January 20, 2020.

Thus, these unproven allegations are not to suggest that CBNRM as a system of governance is problematic or that a substantial overhaul of the system is needed. Instead, it is to suggest that where governance arrangements exist and the oversight of various forms of public funds is managed by individuals or groups, there always remains the possibility for mismanagement and this must be carefully overseen by the state enacting its sovereign power and enforcing mechanisms built into the Nature Conservation Amendment Act with a stringent system of checks and balances. These should be exercised in step with NGOs and global partners to ensure transparency and accountability to accomplish good governance and institutional compliance as outlined in Fig. 5.2. Failing to do so can harm those most in need in society and undermine programme success (Fig. 4.8).

This discussion of allegations of corruption at the local level also demonstrates another way in which CBNRM can only be understood through an analysis of intertwined local and global power and politics. As some of the accounts from Torra described, many of these issues were attributed to climate change, which demonstrates the way the drought has important cascading impacts on the assemblage of actors involved. Moreover, it elucidates the way in which local politics are indivisible from national and global politics in this layered governance arrangement, rather than the community being its own unit of analysis separate from the others. Instead, what we see is the way in which rules and forms of power that exist relationally between actors are indivisible and simultaneously key to determining outcomes for this policy assemblage as will be further unpacked below.

Towards an Empowered Community?

This chapter has unpacked the complex genesis of CBNRM as well as the manifold relations between a multitude of entities and the power that is produced among actors as control is outsourced by the state. Namibia's CBNRM legislation outlines the provision for a devolution of power from the central government to local communities. In this sense communities are able to organize to manage local resources and benefit financially, which often leads to varying degrees of empowerment of local organizations.[122] There is a corresponding degree of authority that is devolved by the state which no longer maintains direct control over the management of these resources or their benefits. This relationship, however, is much

[122] Hoole, "Place-Power-Prognosis."

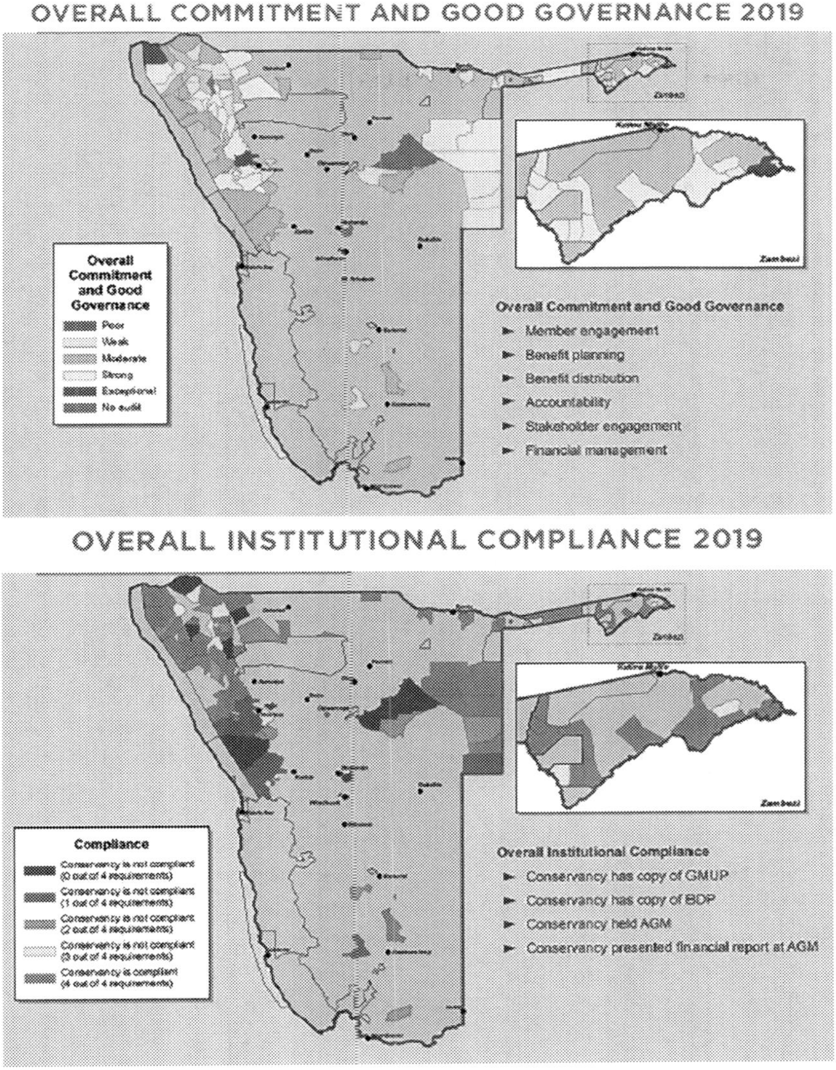

Fig. 4.8 Good governance and institutional compliance (MET/NACSO, "The State of Community Conservation in Namibia: Annual Report 2019: A Review of Communal Conservancies, Community Forests and Other CBNRM Initiatives")

more complex as while conservancies are granted rights to manage certain resources, they are not granted full legal authority over the land itself which is a key piece to the puzzle of wildlife conservation.[123] The success of wildlife is inextricably linked to the effective management of the habitat within which they dwell, and as such managing one without having control over another can result in a number of contradictions for decision making. Thus, communities are given the responsibility of managing resources, downloading costs normally incurred by the state, yet the state maintains mechanisms to exert sovereign power. Furthermore, the Namibian state maintains a number of ongoing oversight mechanisms over the conservancies. The state suggests this is intended to prevent local-level corruption which has and continues to be seen as discussed above, but in practice this coercive oversight also limits conservancies in which activities they can choose to pursue, thus limiting the key pillar of ownership and empowerment.[124] Activities that are limited include the types and quantities of wildlife that can be extracted in a given period which will be further discussed later in the book, what certain areas and types of land may or may not be used for, as well as how resources on that land can be utilized or who will benefit from them including minerals, forests, and freshwater resources. In order to conduct any of these activities or implement changes impacting certain areas, conservancies must apply for and have their request approved by the relevant ministry which can be costly, time consuming, and difficult to achieve.

While conservancies control the rights to 'manage' wildlife this is a bit misleading as the state maintains control over it and issues permits for any form of hunting to be undertaken.[125] This has proven to be controversial as local populations often rely on hunting for meat. Furthermore, in most conservancies trophy hunting has formed the backbone of their tourism operations and can count for well over half of conservancy revenues, while in some as much as 80–90 per cent of revenue at one point was derived from various forms of trophy hunting as seen in Table 4.3.[126] Thus, in times of protracted drought—as Namibia has now been experiencing for almost a decade—far fewer permits are granted and communities' ways of

[123] Brian Jones, Environment and Development Consultant, February 20, 2020.

[124] Interview, Pohamba Shifeta, Minister of Environment and Tourism, January 29 2020.

[125] Boudreaux, "A New Call of the Wild."

[126] Interview, Willie Boonzaaier, Programme Director, Integrated Rural Development and Nature Conservation, January 24, 2020.

Table 4.3 Sources of returns to conservancy members (MET/NACSO, "The State of Community Conservation in Namibia: Annual Report 2019: A Review of Communal Conservancies, Community Forests and Other CBNRM Initiatives")

Year	*Source of cash income and in-kind benefits to conservancies and members*	*Value in N$*	*Percentage of total*
2019	Joint-venture tourism (includes all cash income to conservancies and members)	93,486,012	61.3
2019	Conservation hunting (includes all cash income and meat to conservancies and members)	39,480,595	25.9
2019	Community-based tourism	7,569,429	5.0
2019	Game harvesting for meat, conflict animals, and live sales	6,363,201	4.2
2019	Indigenous plant products	2,155,377	1.4
2019	Miscellaneous (e.g. interest)	1,974,239	1.3
2019	Crafts	1,420,496	0.9
		152,449,349	**100**

life as well as the viability of the CBNRM programmes become increasingly under stress. Nonetheless, this is likely an important and necessary check on power to ensure wildlife populations remain stable and maintain the sustainability of the tourism sector. By devolving power, the state can overcome its inability to govern certain areas and communities are granted the freedom to govern themselves if they do so in a way deemed responsibly by the assemblage of actors involved.[127] In this sense, conservancies are given the responsibilities of quasi-sovereign actors without being afforded many of the tools that often come with sovereignty. This paradox is pointed to either directly or indirectly by much of the literature on CBNRM as well as within the vast majority of the interviews conducted during the fieldwork for this book.[128]

Further complicating these questions around devolution are the tensions between Namibians and SWAPO. As mentioned, Namibia is a country that generally ranks fairly highly among international democracy/

[127] Abrahamsen, "The Power of Partnerships in Global Governance."

[128] Schnegg, "Institutional Multiplexity"; Cruise and Izzy, "Investigation into the Efficacy of Namibia's Wildlife Conservation Model as It Relates to African Elephants (Loxodonta Africana)"; Stasja Koot, Paul Hebinck, and Sian Sullivan, "Science for Success—A Conflict of Interest? Researcher Position and Reflexivity in Socio-Ecological Research for CBNRM in Namibia," *Society & Natural Resources* 0, no. 0 (May 21, 2020): 1–18.

freedom indices, especially in Africa. In 2020 Freedom House gave it a score of 77 out of 100 giving it the ranking of 'free,' based on a number of indicators measuring political rights and civil liberties. Nonetheless, SWAPO has held power for over three decades, despite having experienced a substantial decrease in support with its leader Hage Geingob's share of the vote dropping from 87 per cent in the 2014 general election to 56.3 per cent in 2019 shortly before the fieldwork for this book was carried out.[129] Thus, in theory this is a free and open country which offers regular democratic elections. However, in practice, at least up until the present time, the people have had little hope of realizing a change in leadership, leaving little room to vote for contending visions of land and resource management.

NGOs provide a similarly complex notion of power at the local and global levels. They generally purport to provide resources and technical support and training in order to help communities realize their own goals. Communities, for their part, have become imbricated in CBNRM and conservancy council leaders, staff, as well as conservancy members in general largely responded that they have been empowered by CBNRM and that they maintain decision-making power.[130] While they acknowledge that there are a number of actors who benefit, as well as control and delegate resources within the conservancy, there was a widespread sentiment that CBNRM was supported and managed within the community and that benefits at least partially remained with conservancy members.[131] However, questions remain regarding ongoing activities as individual NGOs have varying priorities which can be seen in the way CBNRM programmes are operationalized in various communities. It becomes difficult to unpack which are local community priorities and which might be indirect spillovers from the will of certain actors within the international community. This is one of the ways analysing this iteration of global environmental governance enables us to examine the degree to which CBNRM activities are developed around creating benefits for local residents in the form of sustainable growth, versus being geared towards

[129] "Namibia: Freedom in the World 2020 Country Report," Freedom House, accessed September 22, 2022.

[130] Focus group 4, Anabeb Conservancy, February 7, 2020.

[131] Interview, Village respondent 3, Uisbasen/Twyfelfontein, January 30, 2020; Farmer 3, #Khoadi !Hoas Conservancy, February 10, 2020; Employee: Grootberg Lodge, #Khoadi !Hoas Conservancy, February 10, 2020.

achieving the goals of global actors such as the environmental movement in the form of the protection of certain wildlife species.

Knowing where decision making stems from can be difficult as NGOs are dependent on international donors. While many of Namibia's NGOs are 'domestic,' almost all receive funding from a variety of international donors. While this is often provided by USAID and similar organizations—supposedly at arm's length to mitigate the need of governments to administer it directly—the interests of donors are shown to impact the types of activities NGOs favour and resultantly discipline the types of conservation activities that take place in various conservancies.[132] The influence of actors who hold the power of global capital can materialize with notions of what conservation should look like, how it ought to be conducted, or even what 'good governance' entails. The current partners providing funding and support for IRDNC as one example include: the Big Lottery Fund, Brot für die Welt, the Conservation Travel Foundation, the Embassy of Finland in Namibia, the Environmental Investment Fund of Namibia, the Global Environment Facility, the German Agency for International Cooperation, the Goldman Environmental Prize, the Houston Zoo, the Minnesota Zoo Foundation, the Namibia Chamber of Environment, Namibia Tours and Safaris, the Open Society Initiative for Southern Africa, TRAFFIC, TUSK, the Umsizi Foundation, the University of Cape Town, the Southern Africa Regional Environmental Programme, the United States Agency for International Development, the Wildlife Conservation Network, the World Wildlife Fund, Mai Family, and the Morby Family Charitable Foundation.[133]

Willie Boonzaaier, programme director for IRDNC, suggests NGOs have been struggling due to core funding being ignored which results in having to spend more time and resources working to procure smaller donations from more donors.[134] Dr Margaret Jacobson, one of the founders of IRDNC, echoed this sentiment lamenting that

> initially those who began the programme largely worked for free and were in the business of humanitarianism, but now it has become big business and increasingly resources go toward procuring resources just to keep the orga-

[132] Boudreaux, "A New Call of the Wild."

[133] "Integrated Rural Development and Nature Conservation - Partners," accessed November 22, 2021.

[134] Interview, Willie Boonzaaier, Programme Director, Integrated Rural Development and Nature Conservation, January 24, 2020.

nizations running and their people employed, with less going toward actual conservation initiatives. This has resulted in too much NGO power, to the point where some organizations see themselves as the programme.[135]

The increasing power of NGOs is another one of the key contradictions of CBNRM, as NGOs are so central to its successful operation that community members at times seem tangential to decision-making processes. NGOs are often staffed with technical experts and well-educated officials whose preconceived notions become difficult not to instil in the management and evolution of projects. Boonzaaier builds on this technocratic rise suggesting that there is an important community empowerment aspect to the programme, but also acknowledges that he sees those NGOs who control funding, having the greatest say and the ability to exert their own will within partnerships. I asked each of the 13 officials I interviewed from various NGOs to what degree their decision making is autonomous from the will of their donors, and each unsurprisingly reported that they make decisions internally within their organizations based on realizing the goals of the CBNRM programme. However, each pointed to the increasingly squeezed nature of funding streams they are constantly in pursuit of, as well as the perpetual requirement to appease both current and prospective donors for their very survival. None of these organizations can exist without donors, and without donors they would not be able to provide services or realize any goals. Beyond the need to endlessly pursue funding streams Cecilia Tortajada argues:

Nongovernmental organizations (NGOs) were once considered as altruistic groups which aim was to impartially influence public policy with no vested interests. Nevertheless, this perception has changed. They are increasingly perceived as groups that prioritize their own ideologies or that respond to the interest of their donors, patrons, and members rather than to those of the groups they represent.[136]

As a result, compromises must be made and rarely do these compromises favour rural African communities. These social and power relations between donor, government, NGO, and community result in a push and

[135] Interview, Dr Margaret Jacobson, co-founder IRDNC, co-director, trustee, mentor, Conservancy Safaris Namibia, Swakopmund, February 17, 2020.

[136] Cecilia Tortajada, "Nongovernmental Organizations and Influence on Global Public Policy," *Asia & the Pacific Policy Studies* 3, no. 2 (2016): 266.

pull as these actors all work to further their interests, though largely working towards similar goals. These complex relationships and negotiations have important impacts on how governance functions, and it is these power relations that the subsequent chapters of this book will further unpack.

Conclusions

This chapter contributes to the discussion of the genesis of CBNRM in Namibia. In it I have built on existing studies that pointed to some of the complex power relations and political manoeuvring involved in the policy that enabled CBNRM. I have outlined the ways in which CBRNM results from a confluence of the concerted actions of a number of heterogeneous actors who are able to exercise their will within the governance arrangement that CBNRM enables. From there, they procure funds from a number of international actors in order to realize certain goals on the ground. As a result, the goals of many are realized at times synergistically, while at others they are in competition with one another, and the interests of certain actors win out over others. This is demonstrated in the activities that occur on the ground and the types of conservation that take place. In order to better understand CBNRM as a form of resource governance and offer meaningful analysis as to its merits for sustainable development and community empowerment, it is first important to understand the history that led to its development.

In many cases, the government has been unwilling or unable to provide important services which are in some instances now being provided by conservancies to ensure the health and well-being of local populations, and so that communities remain safe and accessible for tourists. This is perhaps the quintessential example of the Namibian state and the assemblage of actors using the productive levers of CBNRM to steer subjects in order to provide for themselves the types of services normally provided by government, while realizing important local, national, and global goals. In Namibia there are both public health care and education provided by the state for example, but both important services are subsidized where possible from conservancy benefits accrued through incomes provided indirectly through tourism. Moreover, local tourism operators often contribute directly to public works such as road signs, infrastructure, transportation, and health care. In rural areas far from the capital, public investment in these areas is often ineffective at best, or completely absent at worst. As such, tourism companies will invest in these areas in order to improve community welfare, but also always with their bottom lines at forefront of mind. Without adequate health care they cannot staff their enterprises and without proper roads and signage tourists cannot access their locations or will not want to if the commute is too uncomfortable or precarious.[137] The next chapter will begin examining climate change as a key factor and its impacts on the evolution of this assemblage as well as discuss the challenges it is posing on CBNRM in Namibia.

[137] Interview, Manager, Palmwag Lodge, February 5 2020.

CHAPTER 5

Climate Change: Local Challenges, Global Opportunities?

Namibia has been experiencing a severe and protracted drought for the better part of a decade and while droughts are a normal part of the hydrological cycle, their severity, duration, and frequency are all increasing which is threatening Namibians as well as the flora and fauna across the country. Decreasing precipitation patterns are making life more challenging for everyone and have begun to severely impact the ability of the CBNRM programme to produce benefits that had been enjoyed in the first 15 years of its implementation. This chapter will discuss widespread losses in livestock, veld food for domesticated livestock as well as wildlife, and the increasing pressure this all places on humans and wildlife living in close proximity—this being the essence of CBNRM. The chapter then turns to discuss opportunities that are being provided by some of the very same challenges discussed above. Despite its absence from much of the literature, the chapter demonstrates that people on the ground are acutely aware of the impacts of climate change and are active in efforts to both adapt to and mitigate further harmful effects.

> *Wildlife populations were rebounding all along and they have been hurting the last few years, but it's just because of the drought, not the conservancies.*[1]

[1] Interview, Brian Jones, Environment and Development Consultant, Windhoek, February 20, 2020.

A. Heffernan, *The Global Politics of Local Conservation*, Environmental Politics and Theory,
https://doi.org/10.1007/978-3-031-24177-2_5

CBNRM has been widely studied for its potential benefits as a domestic policy for conservation, sustainable development, and community empowerment. What existing studies have yet to adequately account for, however, is how the impacts of climate change affect CBNRM programmes and their outcomes. As CBNRM is intended to give rights to communities to sustainably manage their local resources in order to commoditize them and benefit economically, any major environmental changes have a direct impact on both the ability to conserve resources and to engage in sustainable livelihoods. While we know the effects of climate change will be felt asymmetrically across the globe, one of the starkest examples of this is Namibia. Namibia is the second most sparsely populated country on earth, with just over 2.6 million people spread across an area of almost 830,000 sq. km.[2] To put this into perspective, neighbouring South Africa's land mass is approximately 50 per cent larger than Namibia's but has a population of about 60 million people—roughly 24 times the population of Namibia. While Namibia is one of the wealthiest countries per capita in Africa, many of its people remain quite poor and with such a small population, it has contributed only a small fraction of the historical global GHG emissions. This discrepancy between a relatively high GDP per capita yet persistent high rates of poverty is a result of Namibia being one of the world's most unequal countries with a Gini coefficient of 59.1, ranking second behind only South Africa according to the World Bank.[3] The fragile climate mixed with high rates of inequality and poverty have important impacts on development initiatives, and CBNRM is no exception. This chapter outlines how environmental impacts are affecting the viability of the CBNRM programme. It also discusses how omission of climate change within policy debates has been detrimental to developmental outcomes as the policies, people, and programmes must evolve and adapt to changing realities in order to continue to garner success. Beyond this, the chapter outlines ways in which climate change has begun to enter into the discussions and activities on the ground—despite being mentioned only rarely in the literature—and certain opportunities have begun being exploited as a result.

The first few years of the current drought resulted in increased discontent with the programme. As the drought has dragged on year after year,

[2] CIA World Factbook, "Africa: Namibia—The World Factbook—Central Intelligence Agency," 2020.

[3] World Bank, "Gini Index (World Bank Estimate) | Data," accessed January 21, 2022.

however, and as the epigraph from Chap. 1 outlines, many Namibians have come to accept that climate change is a very real crisis they will continue to face and that their actions must account for this new reality at every turn.[4] This has materialized in new partnerships at the local, national, and global levels in order to pursue novel funding mechanisms to develop approaches for climate change adaptation and mitigation that can be realized through the governance framework of the CBNRM programme. These partnerships have expanded the types of activities conservancies take part in. They have evolved beyond traditional forms of wildlife management for ecotourism ventures, to now seek to compete to be as green as possible. Conservancies are increasingly working with farmers to develop more sustainable farming techniques, while also finding solutions to protect their livestock and provide water for them, all the while ensuring space, water, and food for the wildlife on which much of their economic well-being is predicated.[5] The CBNRM programme has been evolving in a number of ways that has produced benefits in certain conservancies which have been enjoyed by some members, however, in many cases these evolutions stem from outside the communities as well as outside Namibia. This is seen particularly as NGOs exert significant influence on the direction of the programmes, and as discussed previously NGOs remain largely led by white people who are often not even from Namibia, despite the fact that many have now made Windhoek or Swakopmund their homes for a number of years. Interestingly as partnerships have proliferated, the language of climate change has too. As mentioned in Chap. 3, I did not travel to Namibia with the intentions of seeking climate change as a key object of study, but as the research unfolded, discussions of the changing environment factored into virtually every interview and conversation I had. This demonstrates one of the many strengths of the assembled approach to both data collection and analysis, as it enables us to find what awaits in the field rather than entering with a strictly defined plan and developing analyses to support pre-determined conclusions. These unexpected conversations mirrored many of the activities on the ground that have shifted to fighting climate change, these efforts and the resources that go into them are at times allocated based on who is most well connected to the right players and where the international

[4] Interview, Maxi Pia Louis, Director, Namibia Association of CBNRM Support Organizations, January 14, 2020.

[5] Interview, Pohamba Shifeta, Minister of Environment and Tourism, January 29, 2020.

community will most benefit as a result. The exogenous nature of these evolutions and the issues they can create for local communities will be further unpacked in this chapter.

In some ways these evolutions of CBNRM in response to climate change have resulted in a renewed outlook on the programme as people have shifted from blaming others for newfound challenges and instead have begun to accept their new reality, and act to adapt to it while also working to mitigate further harmful effects. In many cases what these renewed partnerships, activities, and outlooks have resulted in are novel opportunities for CBNRM in Namibia that are slowly leading to diversification which is contributing to making communities more resilient from a climate policy sense, though the degree to which they are becoming successful in developing the third important pillar of CBNRM, community empowerment, remains less clear. Climate change is presenting opportunities for communities as well as the assemblage of actors involved in CBNRM. Opportunities in this sense do not necessarily connote something intrinsically positive, but indicate how adaptation and necessary responses to climate change do provide avenues for some actors to adopt new strategies for profitable enterprises, survival, and pursuing certain interests. Opportunities are presenting themselves in the form of partnerships which are leading to novel activities intended to make the CBNRM programme and the ecotourism activities on which it is based increasingly environmentally sustainable. This includes developing renewable energy systems to power lodges, innovative systems for trapping and reusing grey water, and installing solar systems for mitigating human wildlife conservation. These opportunities bring new actors into the assemblage who come with new funding streams as well as the backing of powerful global organizations. Analysing the local/global politics of CBNRM enables us to account for this constantly shifting set of components as well as their interactions and the outcomes that result. While these evolutions have created important and measurable benefits for the CBNRM programme, they also result in a proliferation of partnerships and a continuing reliance on the will of the international community. This chapter further unpacks these developments in order to better understand the way CBNRM programmes have reacted to their greatest challenge by taking advantage of the opportunities climate change has presented them with. It places a specific focus on the evolution of the actors involved in the programme and the new endeavours being pursued by them, as well as importantly unpacks who in fact is winning and losing as a result of these newfound opportunities.

A Changing Climate in a Fragile Environment

In his 2015 State of the Union Address, then US President Barack Obama proclaimed that climate change is the greatest threat to future generations. Since then, this has become an increasingly incontrovertible statement. Climate change continues to consistently worsen at an increasingly alarming pace. There is ongoing debate as to whether it is being accepted and acknowledged as an existential threat, and for good reason—as collectively the world is continuing to fail to adequately treat it as such. While year after year, improvements are made, and increasing resources and energies are targeted towards important adaptation and mitigation efforts, these remain too few and at far too slow a pace—especially in the Global South. The world is currently on pace to surpass the goal of limiting global warming to 1.5 degrees Celsius above pre-industrial levels set out by the Paris Declaration in 2015. Furthermore, it seems all but assured we will even pass the higher 2-degree increase that was agreed upon by the United Nations Conference of the Parties. Already the effects of climate change are being felt in near-catastrophic ways around the globe, though in asymmetric and perverse ways. This asymmetry is explained by Robert Keohane in asking us to

> imagine two maps of the world One displays the relative severity of environmental problems—air pollution, soil and water degradation, desertification, destruction of habitat—and therefore biodiversity. The other map shows the capabilities that governments have to cope with these problems: the material resources at their disposal, the level of education of their people, the competence and honesty of their governments. Juxtaposing these maps would graphically reveal that environmental problems are most serious in those parts of the world with least capacity to deal autonomously with them.[6]

While it might be assumed this would call for greater immediacy to action, one of the problems is that the world's biggest polluters—both historically and present—are those which are generally set to experience some of the least severe effects of climate change, while simultaneously being in the best position to mitigate these harmful effects.

[6] Robert O. Keohane and Marc A. Levy, *Institutions for Environmental Aid: Pitfalls and Promise*, Global Environmental Accords. Strategies for Sustainability (Cambridge, Mass: MIT Press, 1996), 6.

Sadly, not all countries have this relative luxury, and it is many of the world's poorest countries which are experiencing the impacts of climate change more severely than the global average. This is true for Southern Africa generally, and particularly for Namibia as the driest country in Africa south of the Sahara.[7] Joel Wainwright and Geoff Mann summarize this paradox positing,

> The oft-noted scandal of climate change is that those who caused it will not live to see its full consequences, and those who are suffering or will suffer worst did not cause the problem. This dynamic has a distinct spatial and temporal distribution, through which the living rich enjoy extraordinary privilege relative to the poor and yet to be born.[8]

For Namibia this has resulted in an already dry country becoming increasingly arid year after year and making questions of climate change's existentialism quickly evaporate. The people of Namibia as well as its flora and fauna have spent generations adapting to a harsh environment marked by periodic droughts. However, droughts have been increasing in frequency, severity, and duration for a number of decades compounding the difficulties they must face.[9] Currently they have been experiencing a severe drought that has lasted almost a decade and it is putting increased pressure on virtually every aspect of life in the country. Van Rensburg and Tortajada outline that

> Namibia is the most arid country in Sub-Saharan Africa. In Windhoek, the capital city, accelerated population growth and expanding economic activities, coupled with highly variable rainfall and multiyear droughts, have jeopardized water security and put enormous stress on socioeconomic development.[10]

This has resulted in a plethora of new challenges as well as exacerbated almost every existing issue. For the CBNRM programme—which

[7] Jarkko Saarinen, Political Ecologies and Economies of Tourism Development in Kaokoland, North-West Namibia (Routledge, 2016).

[8] Geoff Mann and Joel Wainwright, Climate Leviathan: A Political Theory of Our Planetary Future (Verso Books, 2020), 72.

[9] Kevin Coldrey and Jane Turpie, "Climate Change Risk Assessment for Namibia's Communal Conservancies," 2020.

[10] van Rensburg and Tortajada, "An Assessment of the 2015–2017 Drought in Windhoek," 1.

intimately ties the well-being of people, to their resources, and thus to local ecological health—these challenges have become starkly evident.[11]

The threat climate change is posing on CBNRM became sufficiently concerning that a report was commissioned by NACSO partners to outline the challenges facing the conservancy programme as well as to provide recommendations for moving forward sustainably. WWF-Norway partnered with other WWF offices including Namibia's to produce research that would help communal conservancies in Namibia adapt to and mitigate the harmful effects of climate change. The report, issued by Anchor Environmental, found that conservancies are experiencing higher population growth rates than the rest of the country and high demands on land mixed with increasingly fragile environments are compounding an already difficult situation. This population growth is predominantly a result of a number of years of difficult financial times for the country, and the hope of benefitting from the services and payments offered by conservancies. Satellite data has identified that in large parts of the Northwest region of Namibia—largely the most CBNRM intensive—especially within conservancy boundaries there have been considerable declines in the productivity of land over the past 17 years. Of the various types of land use, however, wildlife-designated zones tend to be increasing in productivity. This is leading to worries as "there is also increasing concerns that wildlife areas are being reduced through fencing off land for agriculture as well as through invasion by cattle."[12] Thus, the one type of land use that is being demonstrated as environmentally beneficial—wildlife zones—is being encroached upon and reduced by environmentally harmful land use practices—livestock farming. Furthermore, 200 hectares of forest coverage has been lost on average per year since 2001 across northern Namibian conservancies. Forest cover is being lost due to removal for agricultural purposes, use in building and as fuel, as well as through drier conditions that are leading to increased forest fires. Core wildlife zones of conservancies have seen 556 hectares of forest lost and 2209 hectares from multiple use zones between 2001 and 2018.

While it is difficult to isolate exact causes of both land degradation and loss in forest cover, research demonstrates that it is in part due to human

[11] Andrew Heffernan, "Accounting for Climate Change in CBNRM: Reflections on Wildlife Conservation in Namibia," *Journal of Southern African Studies*, 2022.

[12] Coldrey and Turpie, "Climate Change Risk Assessment for Namibia's Communal Conservancies," 2020, vi.

population pressures/activities and changes in the climate. The report by Anchor Environmental uses forecasting tools based on historical averages and suggests total precipitation across Namibia will decrease by nearly 9 per cent for the period between 2040 and 2060 relative to historical (1960–1990) precipitation. This amounts to expected rainfall decreases of 20 per cent in the dry season and 8 per cent in the rainy season. While these patterns are not uniform across the country, mean annual temperature is expected to increase by 3 degrees Celsius (15 per cent)—well in excess of the Paris Agreement's goal of 1.5 degrees Celsius or even the fallback target of 2 degrees Celsius.[13] These increases in temperature and decreases in rainfall are expected in an already extremely hot and dry region where people rely heavily on their local environments for livelihoods and subsistence. As a result of the changing climate, conservancies are being confronted with a number of challenges that are compounding previously existing issues. The findings by Anchor Environmental are corroborated by findings from van Rensburg and Tortajada who present an assessment of the 2015–2017 drought in Windhoek. They outline the severity of the drought which began to be painfully felt by the residents of Windhoek as well as much of the rest of Namibia during this time, but they argue that

> looking back at the data, it is clear that the drought really started with the poor rain and runoff of 2012/2013, yet the effects online became fully visible due to the continuous below average precipitation. Most recently, 2019 saw the lowest recorded rainfall in Windhoek since 1891 and the worst in the last 90 years in Namibia.[14]

This is further outlined in Fig. 5.3 that shows historical rainfall patterns for Namibia. This data demonstrates three severe droughts over the past 50 years which are usually understood to be approximately two-year events. However, as the data demonstrates and as the authors outline above, the precipitation patterns that lead to these serious drought events usually set in well before their effects begin to be felt and last well after, as has been the case with the current drought (Fig. 5.1).

[13] Coldrey and Turpie, vi.

[14] van Rensburg and Tortajada, "An Assessment of the 2015–2017 Drought in Windhoek," 2.

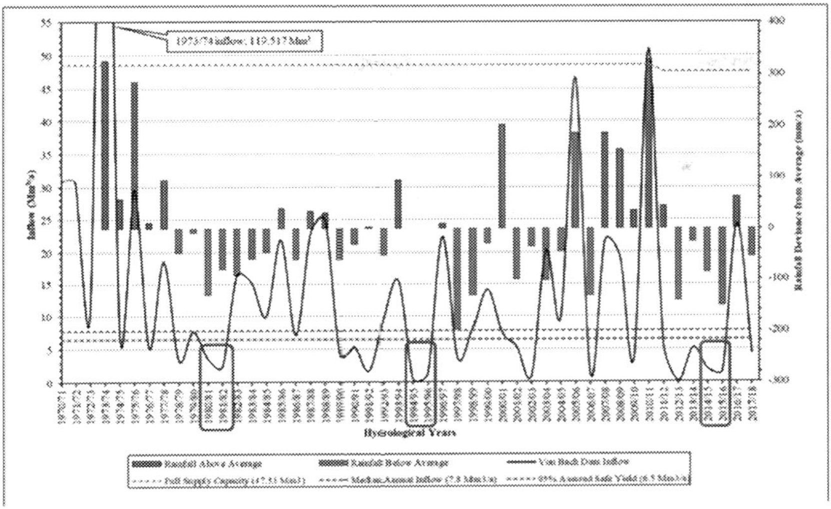

Fig. 5.1 Rainfall variability in Windhoek (Tortajada, "Nongovernmental Organizations and Influence on Global Public Policy.")

As a result of this stark reality, climate change has become a part of our daily lexicon. It is a phenomenon that factors into almost every academic conference, policy strategy, or NGO platform and is pervasive on the global agenda. For Namibians, climate change has become a chronic challenge that is slowly changing the very nature of their landscape and threatening their way of life. This is primarily for two reasons. First, Namibia has been experiencing increasingly erratic climate variability which has demonstrated trends of hotter temperatures and decreasing precipitation levels in what is already the driest country in sub-Saharan Africa.[15] Second, as with much of the rest of the African continent over 60% of Namibians rely on the land to some degree for their livelihoods, and thus shifts in the climate do not simply cause acute disasters, but also chronically limit their ability to feed themselves or to obtain sufficient freshwater.[16] This poses existential challenges as we know "water is a determinant of health and health forms the very basis of human well-being. Water, like health, forms the very basis of human survival and thus the right to life. In many instances

[15] World Bank, "Overview," Text/HTML, World Bank, 2021.
[16] Schnegg and Kiaka, "Subsidized Elephants."

water is both a solution and a threat to human life."[17] For Namibia this has historically been true as its flora and fauna have spent millennia evolving in order to adapt to the dry conditions. However, in recent years water—and specifically its increasing scarcity—has become a growing threat to all forms of life.

In Namibia, discussions of climate change quickly lose their abstract nature and become not only a part of almost every conversation but also enmeshed in most people's every action. The country has been experiencing a severe drought for the better part of a decade and this is widely visible across its landscapes. While droughts are part of the normal hydrological cycle, dozens of interview respondents reported that their severity, frequency, and duration have been increasing in recent decades. These interviews are shored up by stark scientific data that demonstrate how the current drought is beginning to pose existential challenges to some of the regions of the country hardest hit and further disproportionately impacting marginalized populations that are already least well off.[18]

The Namibian landscape is generally a mix of savanna and semi-arid desert with some areas being fully desert.[19] As of 2020 the country was still in the grips of one of the most severe and protracted droughts in recent history.[20] One of the places that present the starkest example of the drought is in the region that is home to one of the country's UNESCO World Heritage Sites—the 'White Lady' Paintings at Brandberg Mountain in Tsiseb Conservancy in the Kunene region. These paintings are estimated to be approximately 5000 years old, and their interest has led to a small village developing nearby where tour guides and their families live.[21] This area, however, has not received rain in over four years which is severely affecting the ability of this village to survive as they dig deeper and deeper boreholes which continuously produce less water. Furthermore, the water is increasingly salinized despite it being approximately 100 kilometres

[17] Ashlea Webber et al., "Water as Threat and Solution: Improving Health Outcomes in Developing Country Contexts," in *Water, Energy, Food and People Across the Global South: 'The Nexus' in an Era of Climate Change*, ed. Larry A. Swatuk and Corrine Cash, International Political Economy Series (Cham: Springer International Publishing, 2018), 187–206.

[18] FAO, "Namibia at a Glance | FAO in Namibia | Food and Agriculture Organization of the United Nations," accessed May 7, 2020.

[19] FAO.

[20] World Bank, "World Bank Climate Change Knowledge Portal," 2020.

[21] Owen-Smith, *An Arid Eden*.

from the coast as the salt seeps into the groundwater once bore holes go deep enough.[22]

Nearby, also in Tsiseb Conservancy is the Brandberg White Lady Lodge—one of the country's most well-known tourist lodges. This lodge provides up to 80 per cent of the conservancy's revenue when it is operating at peak capacity.[23] It is a beautiful lodge nestled at the bottom of the Brandberg Mountain where two rivers—the Uis and Ugab—meet. This confluence of rivers serves an important function for tourism as both the lodge and the campsites overlook this area and during the dry season when water runs short elsewhere elephants, giraffes, oryx, kudu, and other wildlife can usually be seen where these two ephemeral rivers join. However, due to the ongoing drought, the rivers which usually flow for several weeks a year have been dry now for years and the tourism operators and local communities are unsure where the local resident elephants are as they do not know whether there are any remaining sources of water available.[24] The lodge was built specifically for these topographic features, but has in recent years been re-branding increasingly towards the nearby 'White Lady' paintings as the rivers have remained dry with no end in sight.[25] This has become a major part of the ongoing struggle in Namibia as riverbeds are key to life. Particularly during the rainy season when even ephemeral rivers are flowing, it is where communities can procure much of their water, where wildlife spend much of their time, where farmers water their livestock, as well as an important—and at times treacherous—part of transportation in the country.

This key aspect to life for Namibians however has increasingly been under threat as early as 2007 the Intergovernmental Panel on Climate Change (IPCC) was already warning that the impacts of climate change would be more pronounced in the near future than previously thought and research by Rosemary Shikangalah and Benjamin Mapani has found that

> these impacts will have a direct impact on the quantity of water available for domestic and agricultural use as well as induce a systematic decimation of the biophysical environment central to the economic survival of marginal populations living in semi-arid and arid regions of Namibia. The gathering

[22] Interview, Hiking guide, Branberg 'White Lady' paintings, January 28, 2020.

[23] Interview, David Eiseb, manager, Sorris Sorris Conservancy, January 29, 2020.

[24] Interview, Sarah Roos, Manager, Brandberg White Lady Lodge, January 28, 2020.

[25] Interview, Sarah Roos, Manager, Brandberg White Lady Lodge, January 28, 2020.

of social economic indicators from both the Namibia Statistics Agency (NSA) and the Ministry of Agriculture, Water and Land Reform (MAWLR) sheds light on the vulnerability of these marginal populations. This data has been combined together to show how urgent and necessary it is to adapt new management policies so as to sustain populations supported by these ephemeral river systems.[26]

The overall decrease in rainfall demonstrated in Table 6.1 that has been experienced in Namibia as well as the prolonged drought have increasingly become the norm and ephemeral rivers which supply large percentages of the population lie in areas with some of the lowest annual rainfall levels (Table 5.1).[27]

As climate change produces asymmetric impacts between states, it has similarly disproportionate effects within them. Outside of the major cities of Namibia, there have been insufficient investments to build bridges and, as a result, roads tend to traverse riverbeds. When rivers are flowing people generally assess their depth and the strength of the flow of water and—if deemed sufficiently safe—drive through them with their 4 × 4 vehicles. These rivers normally pose one of the major challenges to transportation in the country and one of the considerable dangers—especially to tourists who are less familiar with their strength and depth. However, recent years have increasingly seen dry riverbeds, and while this has facilitated land

Table 5.1 Trends in rainfall 2016–2019[a]

Year	*Khorixas (Station 16)*	*Konop Pos (Station 20)*	*Mopanie (Station 26)*
2019	57.2	102.7	52.5
2018	117.3	141.9	178.3
2017	119.6	166.3	250.8
2016	Incomplete data	Incomplete data	Incomplete data

Tortajada, "Nongovernmental Organizations and Influence on Global Public Policy."

[26] Rosemary N. Shikangalah and Benjamin Mapani, "Ephemeral River Systems and Their Ecosystem Provisions to the Local Populations: A Review of the Huab and Ugab Rivers, Namibia" 14 (May 7, 2021): 47–48.

[27] P. J. Jacobson and K. M. Jacobson, "Hydrologic Controls of Physical and Ecological Processes in Namib Desert Ephemeral Rivers: Implications for Conservation and Management," *Journal of Arid Environments*, Deserts of the World: Namib Desert: 50 years of research in a hyperarid desert, 93 (June 1, 2013): 80–93.

travel for those who can afford vehicles, it is compounding a number of issues for the remaining majority. Predominantly this means society's most marginalized are those who are disproportionately impacted. On the other hand, a lack of rain (up to a point) can perversely benefit wealthy tourists as it makes transportation easier which I experienced first-hand.

With the fieldwork for this book taking place throughout the heart of the rainy season, one of the major and repeated warnings that were offered many times prior to setting out from Windhoek was to exercise extreme caution when driving through riverbed crossings as they can be deeper, have a swifter current, or less solid footing than it might appear. However, throughout the roughly 8000 km covered throughout the research process it was only in Purros Conservancy in the far Northwest of the country that I came across the Hoarusib which was the only river that was actually flowing throughout my time in the country. The dozens of ephemeral riverbeds I either saw or traversed were utterly dry and I was informed most had been for several years now. A further benefit to tourists is that a lack of water makes finding wildlife easier, as they are forced to congregate at the fewer and fewer watering holes. Thus, again while wealthy tourists are more easily able to seek out Namibia's charismatic megafauna for photographic tourism to their delight, this is paralleled by an increasing threat to the lives and livelihoods of local Namibians. To be sure, in areas where the drought has persisted for too long with too much severity, this eventually ceases to be beneficial even for photographic tourism.

For most Namibians, rather than a nuisance for transportation—as the vast majority of black Namibians do not own or have access to vehicles—rivers are treated like gold. This was made clear when the chairperson of Sorris Sorris Conservancy which neighbours Tsiseb to the North proclaimed "if river floods, we are rich" at an event unveiling a new solar plant at the Sorris Sorris Lodge.[28] At this same event one of the area's traditional authorities addressing the crowd half-joked that "the area we are in used to be semi-desert, semi-arid. Now it is desert-desert or arid-arid."[29] This has become less of a joke and instead more of an uncomfortable reality for most Namibians. Even in driving to the event at Sorris Sorris Lodge, I was cautioned that the roadway was very winding, rough, and long, but that due to the drought I could instead easily drive through the riverbed for about 15 km, and it would bring me from the Brandberg White Lady

[28] Interview, Chairperson, Sorris Sorris Conservancy, January 29, 2020.
[29] Interview, Traditional Authority, Sorris Sorris Conservancy, January 29, 2020.

Lodge to Sorris Sorris Lodge, without much difficulty, which proved to be the case. Despite the convenience, it was nonetheless an uncomfortable reality knowing that I was driving through what would have been their gold, and the key to life.

Adolph Thaniseb, the chairperson of Uisbasen/Twyfelfontein Conservancy, slightly further north again from Sorris Sorris, reported that his conservancy has been particularly hard hit and that the "grass here has no energy left in it, so the livestock struggle."[30] The Sorris Sorris Conservancy manager David Eiseb says the country has lost 48,000 livestock due to the drought.[31] This is further corroborated by a Food and Agriculture Organization report that shows findings that as of the first quarter of 2020 "over 59,000 livestock have died due to the ongoing drought situation in Namibia, disrupting the livelihoods of those dependent on livestock and related farming enterprises."[32] David says that they know Namibia is being particularly hard hit by climate change and within Namibia it is the Kunene region that is experiencing the worst of the effects.[33] In similarly asking many rural farmers what the greatest challenges were to the CBNRM programme many agreed with one farmer who responded that "the main issue is drought, so cattle die or lions bite cows and they die. There is a compensation programme set up for this, but the government doesn't give the conservancies enough to dole out, so people are left uncompensated."[34] This quote reporting the fact that livestock are dying as a result of dehydration, while predators are increasingly driven to attack livestock, as the drought has limited their own ability to hunt more traditional prey.

Reading these types of statistics in reports and on websites is always one thing but getting a sense of it first-hand is really something different. When livestock fall dead in a field, they can be useful for a number of purposes and Namibians are careful to quickly process them in order to prevent wasting any possible useful materials from them. As a result, I did not witness dead livestock throughout the fieldwork. Zebras however are being similarly hard hit by the drought, and I had the unfortunate experience of seeing what I would estimate to be approximately 75 dead zebra

[30] Interview, Adolph Thaniseb, Chairperson, Uisbased/Twyfelfontein, January 30, 2020.

[31] Interview, David Eiseb, manager, Sorris Sorris Conservancy, January 29, 2020.

[32] Simone (OCCP) Morini, "FAO Namibia Newsletter, 1st Semester 2020 - Issue #1," 2020, 10.

[33] Interview, David Eiseb, manager, Sorris Sorris Conservancy, January 29, 2020.

[34] Interview, Village respondent 1, Uisbasen/Twyfelfontein, January 30, 2020.

carcasses throughout my time in Namibia. At first, I assumed there must surely be a parasite or disease being passed through the population or that predators were killing them in large numbers. However, like with so many other tragedies in Namibia during the ongoing drought, it was repeatedly confirmed to me that they were dying off due to dehydration. While there have been a number of ongoing efforts between many of the actors involved in CBNRM to develop infrastructure to provide water for wildlife in the field, their efforts are not able to keep up and a number of species are struggling as a result of the impacts of climate change. While it never became easier seeing the seemingly endless carcasses throughout the research, it did in some ways begin to be less surprising. For the first couple weeks of this, I found it hard to wrap my head around the fact that these animals which have evolved over thousands of years in these harsh environments were simply not able to find sufficient water to survive. This is despite the fact that Namibia is home to a specific type of zebra known as Hartman's Mountain Zebra which have evolved to be smaller, require less food and water, and are able to cover more ground than regular zebras. Research conducted by Garth Owen-Smith also found that these zebras can go days and, in some cases, possibly even weeks without drinking water as long as they had access to green vegetation and specific perennial wild grasses on which to graze, making their inability to survive all the more astonishing.[35] However, as the weeks passed and I covered thousands of kilometres of terrain, the one thing that remained constant was a complete and utter lack of visible freshwater sources. Thus, I began to understand their plight and the depth of the interconnected challenges facing both wildlife and human populations.

Local Challenges for Conservancies

As discussions of climate change emerged from the research process, I became interested to find out what people thought about climate change in Namibia. Across the world there are climate deniers as well as those who remain generally unconcerned with the growing threat to global ecological stability and biodiversity. I put this question to Maxi Pia Louis, Director of NACSO, and she responded, "Here people do not have the luxury of not believing in the science. Climate change is all around us, it is

[35] Owen-Smith, *An Arid Eden*.

undeniable, and it is threatening our way of life in the immediate…."[36] As the research progressed, almost every conversation or interview would turn towards climate change, water, and/or the environment. Climate change has direct and deleterious impacts on almost every other component involved in this governance arrangement. Below I will outline some of the ways it is impacting the CBNRM programme as it is having complex and interconnected impacts on the very way governing is able to occur.

The general consensus of my interlocutors was that the most serious specific challenge for the CBNRM programme consists of mitigating issues relating to human-wildlife conflict. These conflicts can stem from increased numbers of wildlife, spreading human settlements that encroach on natural habitat, or by changes in the availability of water and/or food sources for animals. While Africa's charismatic megafauna are beautiful for tourists to observe from the comfort of their lodges or Land Cruisers, many of them are dangerous and destructive for local communities and farmers. In certain instances, predators including lions, leopards, cheetahs, jackals, and hyenas kill livestock, while elephants can stampede through towns and constantly cause costly damage to infrastructure—especially in search of water.[37] This search for water also compounds problems as it makes tracking wildlife easier for those who might wish to do them harm. While poaching has been drastically reduced in recent years as a result of the CBNRM programme, drought increases the likelihood that people will choose to poach, as well as their potential for success.[38] Eric Xaneb, Chairman of Tsiseb Conservancy, says simply "elephants are generally good for tourists, bad for farmers, which is what everyone around here does" (referring to farming) (Figs. 5.2, 5.3 and 5.4).[39]

While CBNRM reduces some of the tensions between farmers and wildlife management, it does not resolve them entirely. This is specifically the case as certain species—such as those named above—are those which are most sought after by tourists and simultaneously those most

[36] Interview, Maxi Pia Louis, Director, Namibia Association of CBNRM Support Organizations, January 14, 2020.

[37] Interview, Michael Knight, Transboundary Leader for KAZA, WWF, January 23, 2020; Interview, Richard Diggle, Business/CBNRM Specialist, WWF, January 24, 2020; Interview, David Eiseb, manager, Sorris Sorris Conservancy, January 29, 2020; Interview, Uisa, Manager, Sesfontein Conservancy, February 6, 2020.

[38] Interview, Maxi Pia Louis, Director, Namibia Association of CBNRM Support Organizations, January 14, 2020.

[39] Interview, Eric Xaneb, Chairman, Tsiseb Conservancy, January 27, 2020.

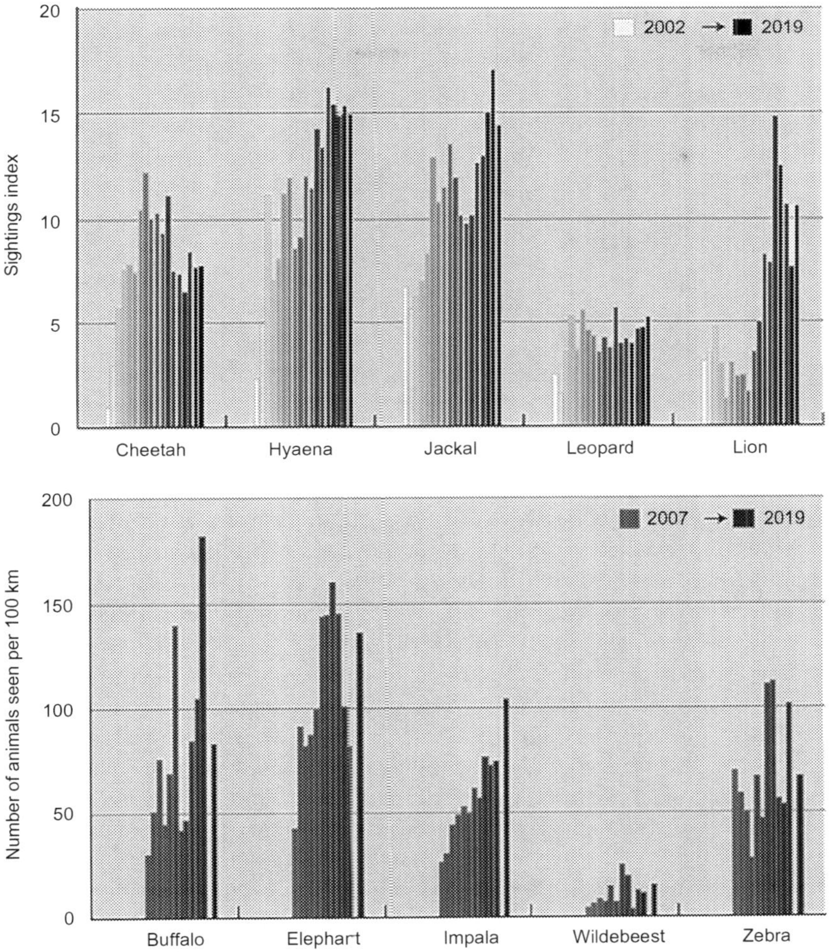

Fig. 5.2 Wildlife sighting trends 2007–2019 (MET/NACSO, "The State of Community Conservation in Namibia: Annual Report 2019: A Review of Communal Conservancies, Community Forests and Other CBNRM Initiatives.") (MET/NACSO)

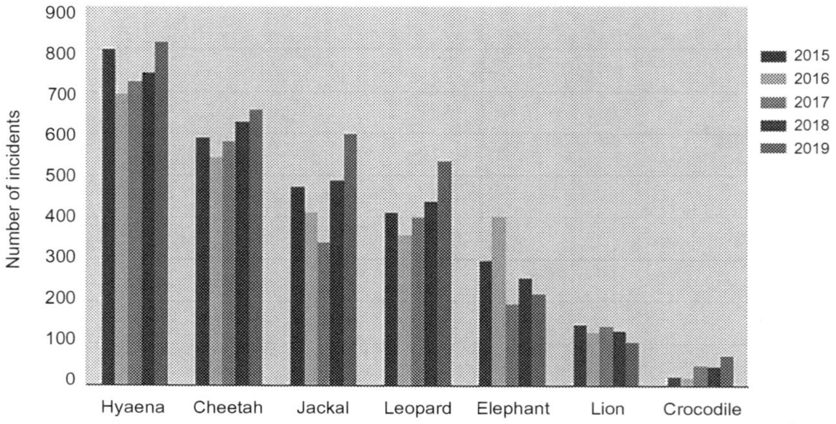

Fig. 5.3 Recorded human-wildlife incidents

threatening to the livelihoods of farmers. Figure 5.5 shows trends in sightings of various wildlife populations and Fig. 5.6 outlines the number of related incidents caused by these. This was one of the major concerns pointed to by some of the sceptics, as many communities initially agreed to implementing CBNRM with the vision of promoting more locally desirable wildlife—specifically regarding safety—such as kudu, oryx, springbok, gemsbok, black faced impala, giraffes, and zebras some of which can be seen in Fig. 5.4.[40] However, what has predominantly occurred is the power of global capital wins out and wildlife that are less locally desirable are often favoured in conservation activities over others. Shikangalah and Mapani demonstrate the types of issues this can lead to as,

> Climate change will also have a direct effect on increasing human-wildlife conflict. With decreasing water resources, lions and elephants tend to access water points in human settlements. While the elephants are usually problematic in the Ugab, in the Huab Catchment, lions have been reported to be problematic (Desert Lion Conservation, 2018). When lions do not find their natural prey (wildlife), they target livestock and as the drought wears on, they become habitual livestock killers (Namibian Association of CBNRM Support Organisations, 2012; Tourism Supporting Conservancy, 2012). In 2018, the lions were reported to have been roaming in the Ugab catchment

[40] Schnegg and Kiaka, "Subsidized Elephants."

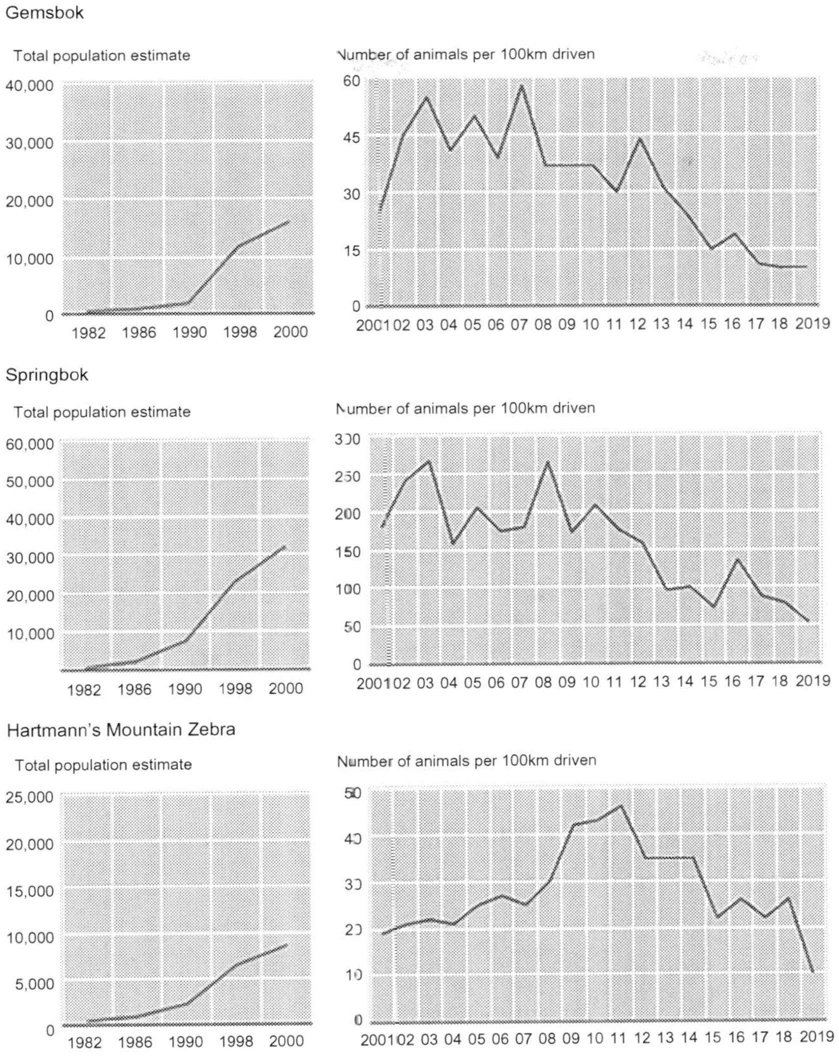

Fig. 5.4 Wildlife population trends (MET/NACSO, "The State of Community Conservation in Namibia: Annual Report 2019: A Review of Communal Conservancies, Community Forests and Other CBNRM Initiatives.")

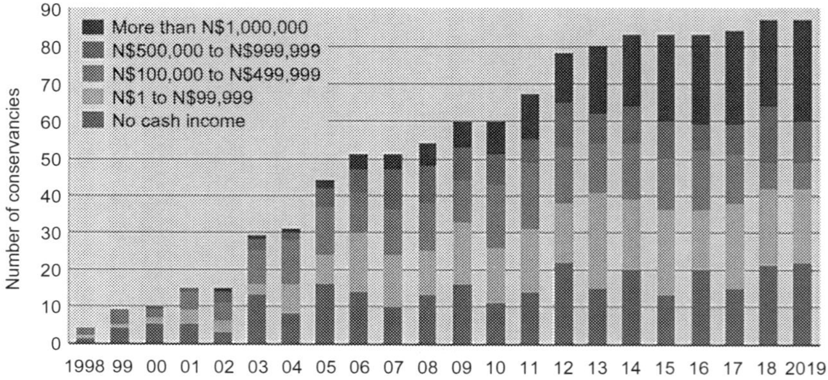

Fig. 5.5 Conservancy cash incomes

> and caused problems associated with killing of livestock (aggressive behaviour) (Desert Lion Conservation, 2018). Preying on livestock is a disadvantage to the local populations, fuelling the human-wildlife conflicts. As such climate change is a major issue that should alert us to adopt methods of protecting these fragile ephemeral river systems for both human and wildlife survival, especially with regard to the abstraction volumes.[41]

Thus, wildlife that are favoured by tourists which can be problematic for communities lead to issues which are aggravated by climate change and the increasing temperatures and subsequent decreasing levels of moisture in the country. Predators are left with less prey to hunt which forces them closer to human settlements they would normally prefer to avoid. One farmer from #Khoadi !Hoas stated, "The biggest challenge we face is drought and because of that we have increased human-wildlife conflict issues so sometimes the wildlife is an issue but it is not because of conservation it is because of drought."[42] Similarly, as elephants fail to find water in the dry riverbeds, they seek out water points meant for either humans or livestock and will rip pipes right out of the ground, both doing costly damage and depleting valuable and increasingly rare freshwater

[41] Shikangalah and Mapani, "Ephemeral River Systems and Their Ecosystem Provisions to the Local Populations," 47–48.

[42] Interview, Farmer 2, #Khoadi !Hoas Conservancy, February 10, 2020.

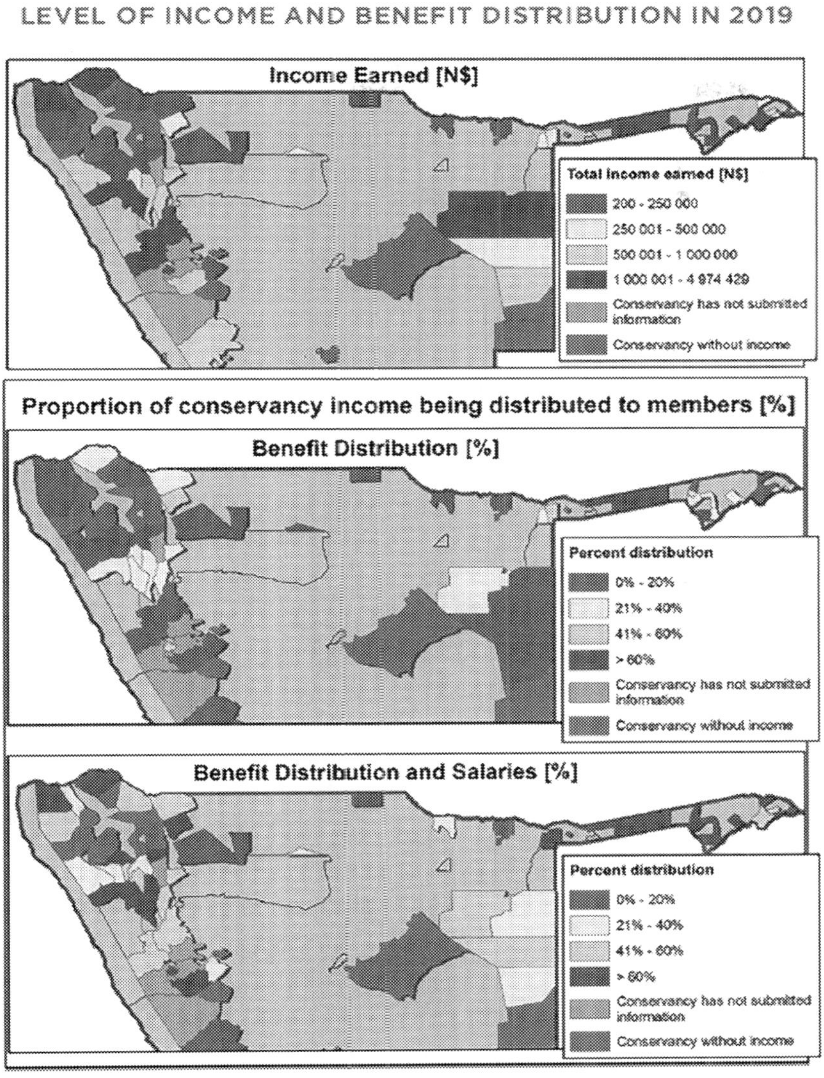

Fig. 5.6 Current state of conservancy incomes (MET/NACSO, "The State of Community Conservation in Namibia: Annual Report 2019: A Review of Communal Conservancies, Community Forests and Other CBNRM Initiatives.")

resources.[43] Next to climate change and the drought, human-wildlife conflict is the subject that comes up in almost every discussion of CBNRM and it is a topic that the many actors involved in programmes are well versed in.[44]

The impacts of climate change are also paradoxical, as decreasing wildlife populations are being paralleled by a rise in human-wildlife conflict. Furthermore, conservancies and their members are experiencing acute economic issues caused by a confluence of factors, but all exacerbated by the effects of climate change. While Namibians increasingly point to conservation as a necessary activity in and of itself, CBNRM has predominantly gained broad support as a result of the financial returns it creates for local residents which can be seen in Figs. 5.7 and 5.8.[45] One MET official with the Directorate of National Parks and Wildlife, Uatirohange Tijuoro, suggested that "drought is the greatest threat to the programme as it makes it seem like its success is diminishing."[46] While exact figures vary a great deal, many respondents point to decreases in direct benefits paid out to conservancy members which has been paralleled by decreasing support of the programmes. While at one time many conservancies were providing each member N$500 per year (approximately US$50), in many cases this has decreased to between N$50 and N$100, while in other cases payments have stopped altogether.[47]

The challenges discussed above are also leading to broader economic issues which have further implications for governance as well. CBNRM developed and has been sustained almost solely around the management of wildlife populations for the purposes of ecotourism.[48] However, since around 2012 there have been a plethora of challenges exacerbated by the ongoing drought that have contributed to these positive results diminishing. While one could argue that ceteris paribus, without climate change

[43] Sem M. Shilongo, Morrie Sam, and Amos Simuela, "Using Incentives as Mitigation Measure for Human Wildlife Conflict Management in Namibia," *International Journal of Scientific and Research Publications (IJSRP)* 8, no. 11 (November 24, 2018).

[44] MET/NACSO, "The State of Community Conservation in Namibia: Annual Report 2019: A Review of Communal Conservancies, Community Forests and Other CBNRM Initiatives."

[45] Stoldt et al., "Transfrontier Conservation Areas and Human-Wildlife Conflict."

[46] Interview, Uatirohange Tjiuoro, Ministry of Environment and Tourism: Directorate of National Parks and Wildlife, January 20, 2020.

[47] Schnegg and Kiaka, "Subsidized Elephants."

[48] Interview, Lara Diez, Nyae Nyae Development Foundation of Namibia, January 10, 2020.

Multiplier effects of industries related to CBNRM include:
- Airlines, hotels and car rental companies:
- Private sector tourism and hunting operations related to conservancies:
- Rental and taxes:
- Further spending generated by the additional income above.

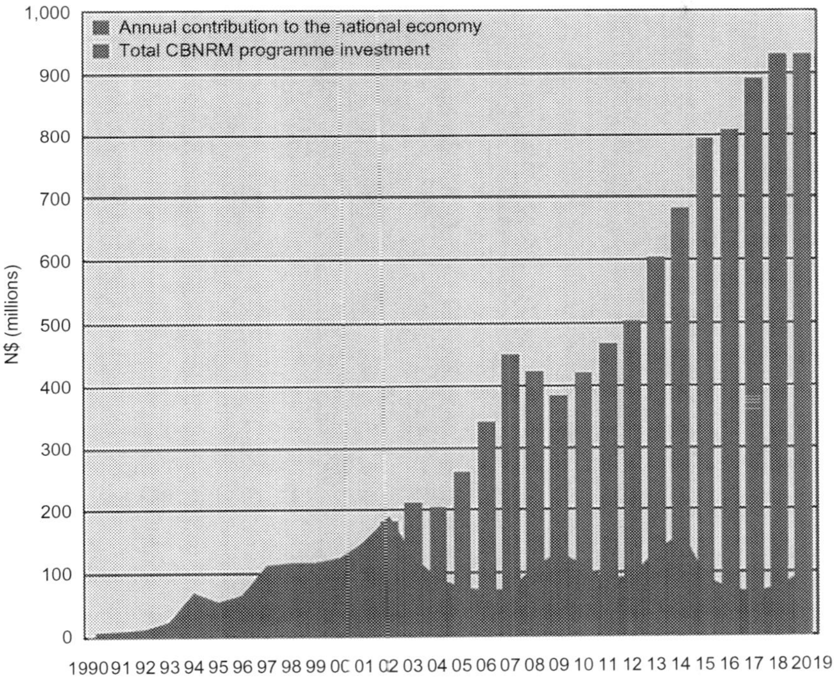

Fig. 5.7 CBNRM national economic returns (MET/NACSO, "The State of Community Conservation in Namibia: Annual Report 2019: A Review of Communal Conservancies, Community Forests and Other CBNRM Initiatives.")

the programme might still be thriving and be set to do so for years to come, such counterfactuals provide little in the way of analytical value. Climate change and every other difficult reality and challenge facing the programme are here and—at least in the case of climate change—are here for the foreseeable future as well as certain to get worse. As such, a system of governance based predominantly on a single resource model that is

Fig. 5.8 Tourism is drought resistant (MET/NACSO, "The State of Community Conservation in Namibia: Annual Report 2019: A Review of Communal Conservancies, Community Forests and Other CBNRM Initiatives.")

already so widely threatened and simultaneously difficult to protect poses a number of challenges to any long-term success.[49]

Over time, as Namibia experienced greater and greater success focusing on wildlife conservation as demonstrated in Fig. 6.1, the programme continuously narrowed in to focus almost solely on that single resource while simultaneously expanding ceaselessly across the country.[50] In this sense more and more of the country's economy and its people's livelihoods have become predicated on an industry based on the governance of a natural resource that is particularly complex and challenging to manage sustainably. In 2020 the tourism sector and related broader economy including indirect effects contributed to over 123,000 jobs (16.4 per cent of total employment) as well as over 11.7 per cent of GDP.[51] This industry being one of the more precarious as it depends first on the global economy and

[49] Interview, Willie Boonzaaier, Programme Director, Integrated Rural Development and Nature Conservation, January 24, 2020.

[50] Interview, John Hazam, Senior Technical Advisor: Namibia Nature Foundation, January 23, 2020.

[51] Namibia Tourism Board, "Namibia Tourism Satellite Account," 2021.

on international tourism which are both subject to volatility,[52] and second it relies on the well-being of wildlife which in turn depend on a plethora of both human and environmental factors. In this sense while CBNRM continues to contribute to certain forms of sustainable development for many Namibians, a growing portion of the economy has also become more dependent on an industry largely predicated on a single resource.

This is understandable as most economies that are able to create consistent revenue streams focusing on the exploitation of a single resource tend to focus their energies most forcefully on it.[53] While CBNRM is only one of many contributors to Namibia's economy, it remains a large and growing one. This has in some ways limited diversification of the Namibian economy and has resulted in structural weaknesses that have been laid bare by the effects of climate change and resultant exacerbation of almost every other existing challenge.[54] A similar process has been experienced more acutely since the onset of the COVID-19 pandemic which Lendelvo, Pinto, and Sullivan argue amounts to a perfect storm that is drastically exacerbating almost every previously existing issue facing the CBNRM programme.[55] The protracted drought has led to less grazing land and veld food, less available freshwater, less prey that is also harder to find, increased human-wildlife conflict, a less hospitable environment for tourists, and lower wildlife populations. All of this has resulted in eroding economic incomes, which has led to a slow erosion of popular support for CBNRM as was demonstrated in much of the work by the sceptics.[56] This existing research was further shored up through a number of my interviews. As John Pallet, Director of the Namibian Environment and Wildlife Society (NEWS), suggested, however, "drought is an opportunity for

[52] Andrew Heffernan, "El discurso oficial de las autoridades marroquíes y sus aliados en la ecologización de sus políticas públicas: modernización del autoritarismo, inserción internacional y lucha contra el cambio climático," *Relaciones Internacionales* 0, no. 42 (September 30, 2019): 157–73.

[53] Macartan Humphreys, Jeffrey Sachs, and Joseph E. Stiglitz, *Escaping the Resource Curse* (Columbia University Press, 2007).

[54] Interview, Lara Diez, Nyae Nyae Development Foundation of Namibia, January 10, 2020.

[55] S. M. Lendelvo, M. Pinto, and S. Sullivan, "A Perfect Storm? The Impact of COVID-19 on Community-Based Conservation in Namibia," *Namibian Journal of the Environment* 4 B (July 1, 2020): 1–15.

[56] Schnegg and Kiaka, "Subsidized Elephants"; Bollig, "Towards an Arid Eden?"

education and evolution."[57] The chapter will now turn to analyse the ways Namibia in general, and the CBNRM programme in particular, is exploiting these opportunities in order to continue to experience success and attempt to build more resilient communities.

Evolving Governance Assemblages

The challenges of climate change are demanding novel approaches to conservation in Namibia and new actors have been seizing the opportunity in order to implicate themselves in the governance arrangement that is ever changing. As a sort of cousin to NACSO, the Environmental Investment Fund of Namibia (EIFN) has emerged to help promote green initiatives across the country. The EIFN developed from the Namibian Government's National Development Plan 1 (NDP1) in 1995 with the mandate to "complement but not substitute for government protection of the environment."[58] The EIFN "was to become a tool for national economic development while securing resources to protect Namibia's fragile environment."[59] Among others, it has established partnerships with the Government of Namibia, the Development Bank of Namibia, the Development Bank of South Africa, the Green Climate Fund, the SADC Centre for Renewable Energy & Energy Efficiency, the Southern African Service Centre for Climate Change and Adaptive Land Management, the Sustainable Development Advisory Council, Namibia Wildlife Resorts, the Global Sustainable Finance Network, the United Nations Development Programme, the Green Fund, South South North, the Desert Research Foundation of Namibia, and the Namibia Energy Institute.[60] These are just the major players, but this list alone demonstrates the political capital behind this network as the key development, environmental, and governance entities in Namibia are linked to a number of most powerful similar such entities on the global stage with the EIFN as the arm of the state which serves as an interlocutor to this assemblage of actors. Duffy argues these are part of a growing and powerful complex global network of governments, international organizations, and environmental NGOs that has

[57] Interview, John Pallet, Director, Namibia Environment and Wildlife Society, Windhoek, February 21, 2020.

[58] EIFN, "Our History: EIF Through the Years," n.d.

[59] EIFN.

[60] EIFN, "Partners," accessed November 23, 2021.

been taking conservation out of the hands of certain states—predominantly those in the Global South—and placing it into the hands of non-state entities and complex non-territorial networks of governance.[61] Resultantly there are clear connections that are established in order to work towards achieving sustainable development goals.

The energies of these organizations are more comprehensively focused on climate and the environment and have further disciplined efforts that were already underway domestically to improve the sustainability of CBNRM programmes. As the Namibian organizations listed above have realized opportunities to acquire funding from their global counterparts, they have sought to work with conservancies in order to maximize greening efforts to become increasingly appealing for donor funding. This is part of the strengths of the way CBNRM has emerged in Namibia as it is a system for managing wildlife, but at its core it is a system for democratic governance, and the many actors that enable it can work synergically towards conservation and resource management activities. While some point to the increased layer of governance that can simply result in another cumbersome bureaucratic institution,[62] conservancy councils have provided a form of democratic representation at the local level which did not previously exist in some cases, or enshrines the territory's governance by traditional authorities in others.[63] This representation has helped enable the realization of a number of positive outcomes in conservancies. The recent partnerships that these councils have been engaging in can also demonstrate that global conservation schemes are highly political interventions that produce new forms of conflict centred on control of and access to natural resources.[64] This conflict can occur as the Namibian state devolves control over wildlife to communities, yet maintains oversight over them as well as control over land use allocations. Africa's rare charismatic megafauna are considered a global good by many and thus many outside of Namibia feel it is within global public interests to protect them, while others also wish to travel to see them. All of this creates a complex picture of tensions and interests which generally desire to see the

[61] Duffy, "The Potential and Pitfalls of Global Environmental Governance," 98.

[62] Cruise and Izzy, "Investigation into the Efficacy of Namibia's Wildlife Conservation Model as It Relates to African Elephants (Loxodonta Africana)."

[63] Kalvelage et al., "Territorialising Conservation."

[64] Duffy, "The Potential and Pitfalls of Global Environmental Governance."

protection of this important natural resource, but each pursuing their individual interests and leading to forms of competition.

For Namibian conservancies, these competing interests result in complex and largely divergent outcomes. While virtually every conservancy member I spoke to suggested the CBNRM programme had resulted in improved representation and being given a voice where previously they had largely not had one, it is clear that having a voice does not always mean being able to affect outcomes. Conservancy members feel they are given voice as the CBNRM programme as a system of political organization has enabled the democratic election of councils to represent their people. While these systems of governance are not simply filling a space which had previously been a vacuum, the existing systems had been far more informal, less democratic, and had limited connectivity to outside actors than conservancy councils have become.[65] As the programme is already so prolific throughout the country, it has become an effective inroad for the delivery of new initiatives. Previously outside actors were not connecting with traditional authorities who held power in rural Namibian communities, nor did these leaders have an effective avenue for seeking this type of funding. The CBNRM programme has provided these types of avenues, as NGOs have partnered with conservancies and conservancy councils have become enmeshed in an increasingly dense network of actors working towards sustainable development goals.

Above I have discussed some of the ways in which some of the energy had been dissipating from the CBNRM programme over the years. This was something that was reported in a number of interviews and something I could visibly see at times. Joglinde Touros, the manager of Twyfelfontein/Uisbasen Conservancy, made this sentiment abundantly clear during the conservancy's official annual audit when the floor was opened for comments, and she stated,

> So many of us have lost why the conservancy programme was established. There is nothing happening outside of the offices… Sure it is good to learn governance, but land is going, water is going, trees are dying, but we are not talking enough about these things that matter.[66]

[65] "Traditional Authorities Act 25 of 2000 - Namibia Legal Database," 2000.

[66] Interview, Joglinde Touros, Manager, Uisbasen/Twyfelfontein Conservancy, January 30, 2020.

Part of this loss of energy seems attributable to the natural progression of a programme that has been around for a long time, that becomes ingrained in the system and partially taken for granted. While novel development initiatives which promise grand outcomes can garner a great deal of resources and the talents of ambitious people, if expectations fail to be met often resources and energy will dissipate over time. Part of this dissipation is evident in the very people who formulate integral parts of the governance assemblage. During the 1980s when the call for sustainable development based on conservation grew partly as a result of the work of Garth Owen-Smith who is widely credited with 'founding' the CBNRM programme in Namibia, many young, energetic people gravitated towards the cause and poured themselves into it while getting very little in return. Over time as CBNRM developed as an important national industry, remuneration improved, and an entire network of NGO staff and officials emerged predominantly in Windhoek. Now 30 years on, the programme seems at a precipice where most of the same people are heading towards the tail ends of their careers, which means that most of those leading NGOs distributing resources and providing training to conservancies tend to no longer be young and have naturally lost some of their youthful energy and idealism. Sadly, this transition phase seems to have been symbolized with the unfortunate passing of Garth Owen Smith in April 2020 shortly after the fieldwork for this book was conducted.[67]

Namibia's CBNRM programme must now begin to deal with the tough questions of needing to replace its founding generation, while much of the next generation lacks education, training, and experience as well as the institutional capabilities for effectively managing certain areas of the programme.[68] These issues are compounding the challenges posed by climate change which have been elaborated above. Several interviewees pointed to this as being a growing issue while also suggesting a sort of changing of the guard was slowly occurring in which the older white people who had largely founded the programme were being replaced by young, black Namibians.[69] This provides both challenges and opportunities and this transition will be key to the future success of the programme. While

[67] IRDNC, "Tribute to Garth Owen-Smith, African Conservationist (Commentary)," Mongabay Environmental News, 2020

[68] Dr Ekkehard Klingelhoeffer, Head of Department—Wildlife Management and Ecotourism, University of Namibia, January 13, 2020.

[69] Interview, Dr Margaret Jacobson, co-founder IRDNC, co-director, trustee, mentor, Conservancy Safaris Namibia, Swakopmund, February 17, 2020; Interview, John Hazam, Senior Technical Advisor: Namibia Nature Foundation, January 23, 2020; Interview, John Pallet, Director, Namibia Environment and Wildlife Society, Windhoek, February 21, 2020.

ownership—as mentioned several times throughout this book—has always been a central pillar to the programme and most involved desire to see young black Namibians take over, they are not always formally educated or as experienced as those who came before them. A lack of formal education and experience was one of the key limiting factors pointed to in almost all of my interviews regarding the programme.[70] Professor Ekkehard Klingelhoeffer, head of the department of Wildlife Management and Ecotourism at the University of Namibia, further suggested that this was one of the biggest challenges of the programme, as NGOs want to turn the projects over to local people but that they are not sufficiently educated in financial matters and business management.[71] This was discussed in Chap. 4 due to the compound lasting effects of both colonialism and Apartheid Rule that undermined every aspect of development for Namibians for decades. While a great deal has been achieved over 30 years of independence, the scars of colonialism run deep, and inequality and poverty remain persistent problems.

Despite a number of complex challenges associated with this changing of the guard, it is also likely an important and necessary evolution in order for Namibians to continue to develop the third pillar of CBNRM and attain ownership over their development. While Namibians at the local level have gained increased representation and voice as a resultant of conservancies and conservancy councils, NGOs and their officials continue to have a disproportionate impact on steering outcomes, and as has been discussed, these NGOs have largely been staffed by white and often non-Namibians. While a lack of education and experience was often pointed to, more Namibians are being educated at the elementary, secondary, and post-secondary every year as well as gaining important experience in relevant areas. The ongoing challenge of a dearth of experience for many is the same kind of challenge many university graduates in the Global North face when they leave the halls of education and seek employment but are told they lack experience. Now is the opportunity for Namibians to take over these roles and both gain the experience needed to continue the work of the assemblage of NGOs and NACSO partners. Beyond this, however,

[70] Interview, Uatirohange Tjiuoro, Ministry of Environment and Tourism: Directorate of National Parks and Wildlife, January 20, 2020.

[71] Interview, Dr Ekkehard Klingelhoeffer, Head of Department—Wildlife Management and Ecotourism, University of Namibia, January 13, 2020.

it will be just as important for Namibians to put their own historical and cultural experiences into the NGOs in order to instil local nuance and continue to move away from cookie-cutter development approaches. Obviously the notion of empowering Namibians is not novel, and it is also something that has been occurring for years, with many black Namibians filling staff positions with a number of NGOs. However, the majority of director and executive positions remain with legacy white officials who worked hard to get these organizations up and running and can be accredited with much of the success CBNRM has delivered to date. One of the cornerstones of CBNRM has always been for conservancies to become self-sustaining. While this book has outlined the ongoing need of NGO technical support on the ground, a next crucial step in the programme's evolution would be for Namibians to be in positions of power with these NGOs in order to be the ones developing this experience and providing technical assistance where it is needed.

Interestingly, the challenges of climate change are creating opportunities that are assisting in providing new initiatives which are helping with this transition. As the impacts of climate change have worsened, many of the efforts of CBNRM have shifted to developing innovative practices to increase the sustainability of the programme which at times have served to simultaneously treat some of Africa's ongoing long-term challenges.

One of these challenges that Africa faces as a continent is access to reliable, clean power.[72] This is a particularly difficult challenge for a country like Namibia that has both a small and an extremely sparse population. The small population means that total GDP and hence investment and purchasing power are low. Having a sparse population compounds this as any investment that is made in infrastructure—such as for power generation—is spread out across fewer people and greater distances, vastly increasing gross as well as per capita costs. Namibia is home to vast expanses of territory with extremely low-density populations. With settlements spread out like this it is difficult to develop power-generating stations that are economical and transporting that power so far, for so few, becomes a complex and costly challenge. With the economy developing around an industry like tourism, having a reliable power source becomes essential.

[72] Kristine Bos, Duncan Chaplin, and Arif Mamun, "Benefits and Challenges of Expanding Grid Electricity in Africa: A Review of Rigorous Evidence on Household Impacts in Developing Countries," *Energy for Sustainable Development* 44 (June 1, 2018): 64–77.

Beyond CBNRM and ecotourism, scholars have long posited the importance of reliable, sustainable power as a necessary cornerstone for effective development for Africa.[73] However, as many rural Namibians do not have such infrastructure in place, many towns, tourist lodges, and campgrounds have procured diesel generators to ensure that there is a consistent source of power. Even for those who are on the electricity grid where it exists, generators are often a necessity as these tourist venues can use a substantial amount of power, and tourists paying top dollar do not take well to outages.[74] Generally this means that people either got by without power at times or invest in expensive generators that ceaselessly burn costly and dirty diesel fuel. This is far from an ideal solution as diesel is both expensive and difficult to transport to remote areas and generators are prone to breakdowns and produce a great deal of GHG emissions as well as noise—not a welcome addition to a vacation in the serene Namibian savannah.[75]

To treat this issue, one of the major efforts that have been underway across the country is to harness the one resource that Namibia has a relentless wealth of—sunlight. Although investing in solar energy may seem like an obvious approach for a country like Namibia, it is not such a simple equation. Even in locations where the sun burns bright most days of the year, the investment costs can be prohibitive—especially with a GDP as small as Namibia's.[76] Between issues of cost, infrastructure, transportation, and others, solar power has not widely been harnessed in Namibia as well as many other countries in the Global South. As the realities of climate change have increasingly brought the world to action however, and as its effects have been continuously impacting Namibia, the network of actors discussed in the previous section have been working to develop solar generating capacity in targeted areas across the country.

One particular example of this occurred on January 29, 2020, when Sorris Sorris held the official launch event for one example of what these partnerships can accomplish in working with the CBNRM programme.

[73] Helene Ahlborg et al., "Provision of Electricity to African Households: The Importance of Democracy and Institutional Quality," *Energy Policy* 87 (December 1, 2015): 125–35.

[74] Heffernan, "El discurso oficial de las autoridades marroquíes y sus aliados en la ecologización de sus políticas públicas."

[75] Interview, Benedict Libanda, CEO, Environmental Investment fund of Namibia, January 29, 2020.

[76] Heffernan, "El discurso oficial de las autoridades marroquíes y sus aliados en la ecologización de sus políticas públicas."

Sorris Sorris is approximately 350 km Northwest of Windhoek and has about 345 registered members. It is one of the original conservancies, having been gazetted in 1998. Sorris Sorris translates to something like 'sun and sand' which is very evident by the landscapes and topography as it is a particularly dry, desertic region of the country.[77] The conservancy's main tourist attraction, Sorris Sorris Lodge, is a breathtakingly beautiful upscale lodge that costs an average of around US$700 per person per night—though it can go well upward of this depending on the type of package. This lodge is extremely remote and there are no towns in close proximity. As such, it is completely off the power grid and since the time it was built in the late 1990s has run on a large diesel generator. This generator has been burning an average of 40,000 litres of diesel per month which cut into profits that are shared between the lodge owners and the conservancy. Over the past two decades it has also contributed a great deal in the way of GHG emissions, further compounding the environmental problems faced by CBNRM, Namibians, and the broader planet. Through a complex assemblage of partnerships however, the EIFN worked closely with Sorris Sorris Conservancy and the owner of the lodge to procure funds from the Green Climate Fund donated by 49 countries, regions, and cities globally to invest in a solar plant that will power the entire lodge at peak capacity.[78] This project resulted from roughly three years of planning and two years of construction. Materials were sourced locally when possible and brought in slowly by 4 × 4 vehicle to navigate the extremely rugged terrain to get to the lodge.

Most importantly the new solar plant immediately prevents a great deal of carbon being burned into the atmosphere. This point was made central at the ceremony for the unveiling of the power plant, when David, the conservancy manager, paused his speech and slowly and powerfully stated, "No more diesel… No more diesel!"[79] Beyond this, the partnership procured the funds to build the solar plant specifically for the Sorris Sorris Conservancy and it is thus entirely owned by the conservancy members. Therefore, the lodge now purchases the power from the conservancy and the funds go directly into conservancy activities and towards benefits for

[77] Interview, Benedict Libanda, CEO, Environmental Investment fund of Namibia, January 29, 2020.

[78] Green Climate Fund, "Initial Resource Mobilisation," Text, Green Climate Fund (Green Climate Fund, October 26, 2021).

[79] Interview, David, manager, Sorris Sorris Conservancy, January 29, 2020.

local residents in surrounding communities that are located approximately within a 20 and 70 km radius. This creates a constant flow of much needed cash to support local communities while also providing the lodge with a quieter source of reliable renewable energy.

The conservancy members I spoke with suggested this means much more than money in their pocket in the short term. Instead, they spoke predominantly of the long-term goals this investment would allow them to achieve. Up until this point a great deal of medium- to high-skilled positions within CBNRM had been filled by NGO staff who often came from elsewhere and were predominantly white as discussed above. One of the major efforts in Namibia, however, has been to educate and train its black populations so they are able to take over the management of CBNRM activities. David stated that beyond realizing important climate goals, the funds from the ongoing sale of power to the lodge would enable the conservancy to create a scholarship fund to send young girls off to the University of Namibia to become educated in fields of study such as wildlife conservation and solar power installation and maintenance so that in the future, they would not need to rely on this type of international partnership to achieve their sustainable development goals.[80] This is one of the ways that climate change is also producing opportunities for a number of actors within the assemblage. It is enabling the GCF and its donors to point to both climate and sustainable development results, while the Namibian state can also claim similar victories while having served more as a middleman rather than having provided any of the resources or services required to achieve these ends.

Beyond these immediate economic and climate benefits, the solar plant is expected to serve as an extra draw to tourists, as many wealthy people from the West travel to Namibia as 'ecotourists' and have increasingly been seeking out the most sustainable lodges and tourist activities. It is easy to see why this would provide Sorris Sorris Lodge with a competitive advantage, as the solar plant has a particular beauty all its own—especially for the environmentally minded. Besides its visual intrigue, the story behind the network that procured its funding as well as the technology and innovation that went into building this high-tech system in such a remote location in the desert that is not even serviced by what most would consider a road is a considerable success. One can therefore easily imagine that when tourists are choosing between the many lodges in the Kunene

[80] Interview, David, manager, Sorris Sorris Conservancy, January 29, 2020.

region this solar plant would likely give Sorris Sorris Lodge a leg up on the competition. This is being increasingly reflected as lodges compete against one another to become ever greener. One of the oldest, most luxurious, and well-known lodges in Namibia is the Damaraland Camp in Torra Conservancy and its manager reported that the entire lodge can be removed in two days and leave absolutely minimal footprint behind.[81] This fact has served as an important marketing mechanism for the lodge for years and it remains one of the most profitable in the country. It also contributed to it being nominated in 1998 by the British Guild of Travel Writers as one of the top three ecotourism destinations on the planet.[82] Therefore, investments like the solar plant at Sorris Sorris Lodge can have many positive economic offshoots beyond the obvious and important climate benefits.

Beyond the potential benefits for sustainable development the solar plant can bring to Sorris Sorris, the politics of it are both interesting and simultaneously remain opaque. As elaborated above, being awarded this type of funding to build such a solar plant would generally be desired by all conservancies—both in Namibia and elsewhere. So how is Sorris Sorris came to be awarded such an endowment? The answer to this question remains somewhat muddied, but it appears that Sorris Sorris had the right mix of having one of the higher end lodges in the country, owned by a well-connected investor, being geographically situated close enough to Windhoek, as well as several of the country's more famous tourist sites. Furthermore, both the conservancy leader and council as well as the area's traditional authority are well-connected within the right circles. The conservancy manager for instance previously worked with the Director of NACSO, Maxi Pia Louis, who also previously sat on the EIFN board with its current CEO, Benedict Libanda.[83] Maxi Pia Louis also has a close relationship with the Minister of Environment and Tourism that began prior to independence as the two were closely involved in the—at times violent—resistance to Apartheid South Africa. They have since continued to work closely together on CBNRM as two of its central figures in the country.[84] This group of actors worked with the EIFN and its partners, NACSO

[81] Interview, Manager, Damaraland Camp, February 2, 2020.

[82] Owen-Smith, *An Arid Eden*, 533.

[83] Interview, Maxi Pia Louis, Director, Namibia Association of CBNRM Support Organizations, January 14, 2020.

[84] Interview, Pohamba Shifeta, Minister of Environment and Tourism, January 29, 2020.

and its affiliate NGOs, and all levels of government in Namibia to put forth the successful application to the GCF to be awarded the resources for the solar plant. While it might seem obvious that the more well-connected will garner greater success in these types of endeavours, it is nonetheless important to acknowledge these political realities so that questions are asked as to whether this is the most efficient or effective allocation of scarce development resources.

Libanda is a key cog in the application process for this type of funding and believes appeasing the global community is integral to success in attaining grants. For Sorris Sorris, at least one part of this appears to be the fact that the conservancy has made a conscious and public decision to interdict trophy hunting.[85] This unveils an interesting way politicking occurs in relation to CBNRM as certain global actors allocate rewards based in part on the types of activities they support or avoid—in this case trophy hunting as will be further unpacked in the next chapter. This is also a way in which productive power can be seen throughout the CBNRM programme, as Western science regarding climate change, conservation, and wildlife populations is presented as truth, and those who abide by it, and by the desires of the West, tend to be rewarded—albeit often indirectly—for their actions. However, in discussing the issue further with David Eiseb, he reports that one of the main reasons Sorris Sorris Conservancy is not a proponent of hunting is because as their conservancy is so desertic, they have modest wildlife populations to begin with and hunting would neither be particularly lucrative for them nor be sustainable. Thus, unlike some other conservancies they must do all they can to protect the limited wildlife numbers that do reside within their boundaries.[86]

Even the structure of the assemblage of actors provides an interesting example of the way these types of partnerships have evolved. While at one time aid tended to be channelled through bilateral government to government relationships, or via global multilateral institutions and NGOs, increasingly we are seeing the proliferation of private and hybrid foundations and funds that are changing the development landscape. This has resulted in a proliferation of donors, and more types of donors that are not constrained by the usual types of democratic and bureaucratic processes which can be cumbersome, but which are also integral to effective

[85] Interview, Benedict Libanda, CEO, Environmental Investment fund of Namibia, January 29, 2020.

[86] Interview, David Eiseb, manager, Sorris Sorris Conservancy, January 29, 2020.

governance. It continues to remain unclear what precisely the long-term outcomes of these shifts will be, but in the short term what we have seen is a disembedding of sorts of aid from democracy. Jean-Michel Severino and Olivier Ray suggest this de-democratization of aid has been proliferating over the past several decades as there has been a marked shift away from official development assistance, to various forms of aid provided by a variety of non-state actors.[87] For them and others, this has served to mitigate certain previously existing issues, while simultaneously opening a number of avenues for funding to be more effectively targeted to where it is needed. In some ways, however, it has also removed some of the politics of development funding which can lead to questions of transparency and how decision making takes place. What is seen in practice however with organizations such as the GCF is a more complex, opaque, and technocratic process that leads to the many critiques discussed above. While most states are held to account by their citizens, and—at least in theory—must account for where, how, and why funds are allocated, private interests do not have this same level of responsibility. This is because it is most often shareholders and stakeholders they are primarily beholden to rather than through transparent, public, and democratic decision-making processes. As a result, private interests tend to work only for their own bottom line, rather than for broader social uplift as is at least the rhetoric of most democratic state public development efforts. As public and private become increasingly blurred, and as governance is devolved by the state through initiatives such as CBNRM, this can often result in private interests developing an increasing foothold in governance which provides them with a growing degree of influence.

The solar plant application for Sorris Sorris was years in the making and took specialized technical skills and political connections in order to be successful, as well as a baseline level of knowledge, skill, and resources from the conservancy council and its members in order to demonstrate their deserving of the grant. Sorris Sorris was one of the earliest conservancies established and as an actor had developed the types of capacities needed, in order to manoeuvre within the assemblage of actors to ply resources from the international community to contribute to its welfare. These capacities were developed in part by the conservancy being gazetted early, quickly becoming economically self-sufficient and producing

[87] Jean-Michel Severino and Olivier Ray, "The End of ODA: Death and Rebirth of a Global Public Policy—Working Paper 167," Center For Global Development, 2009.

community benefits that among other things contributed to scholarships to send their young residents to attain post-secondary educations. These are ways in which the conservancy was conditioned along certain lines in order to become increasingly responsibilized for its own outcomes—though outcomes that followed a largely pre-determined path.

In and of itself, it is not inherently problematic that those most well-connected or conservancies who interdict trophy hunting are awarded in this way by the international community. However, it begs the question as to whether funding is being allocated based on where it is most needed and/or will have the most beneficial impact. Furthermore, it risks leveraging existing power imbalances in order to see the power of global capital sway conservancies down a particular development trajectory which creates prospects for continuing the same type of neocolonial cookie-cutter solutions to development which have long track records of demonstrated failure. Favouring one conservancy's approach to conservation activities over another risks ignoring factors such as what type and quantity of wildlife they might possess which can also be the case with any other type of resource. This is not to argue for a re-centralization of governance, but instead to acknowledge the heterogeneity of communities in all its forms. This means accounting for ethnic, cultural, political, historic, and natural differences (i.e. accounting for the local environment and natural resources). Natural endowments greatly impact the types of conservation that are possible, as well as resources that can be sustainably profitable. These are the types of considerations that ought to be carefully analysed in awarding development funding. However, it seems that one potential consequence of the aforementioned disembedding of development assistance from more traditional forms of Weberian democracy is that there is often less transparency and accountability as to how funds are allocated than through more traditional public channels. Instead, what emerges are novel democratic processes that occur outside of the traditional state halls of power. Since the failures of structural adjustment in the 1980s and 1990s development scholars and practitioners have long posited the need to move away from cookie-cutter solutions in order to account for local nuances. CBNRM provides opportunities for additional autonomy for conservancies to make their own decisions based on local realities, and both these decisions and factors that are accounted for in the decision-making process ought to be accounted for when reacting to them and allocating resources as a result. This is specifically another reason why it is integral that Namibians begin to fill positions of power for NGOs that make decisions and allocate such resources.

These are the types of social relations that the CBNRM programme both creates and enables through the devolution of power. The outcome is often an arena in which space emerges for a multitude of actors to engage in efforts to pursue their own interests and attempt to realize their own ends. While the example of the Sorris Sorris power plant can be said to be a success story, there also remain difficult questions around who these funds are intended to benefit and who in fact is made better off as a result. While here I argue that the power plant has resulted in a number of winners in the sense that the lodge becomes more sustainable and profitable, the global community benefits from less GHG emissions, the tourism of wealthy westerners becomes more sustainable, and local communities also benefit from cleaner air and sustainable profits from the power produced, this seems to create a number of benefits in one of the conservancies which was already most successful which does little to benefit those in other communities. These tensions will be further unpacked in the next chapter.

Among the new partners outlined at the beginning of this section, the Green Climate Fund is perhaps the most prolific, interesting, and simultaneously one of the more controversial of the organizations. The GCF was conceived of at the 2009 United Nations Framework Convention on Climate Change (UNFCCC) Conference of the Parties (COP) 15 in Copenhagen with the promises to channel $100 billion annually to scale innovation and counter climate change with a "paradigm shift towards low-carbon and climate resilient development pathways."[88] Its purpose is to join environmental and development policies to realize these global goals synergistically. The GCF was intended to entail a paradigm shift that would "grant or lend to both the public and private sector simultaneously; while also generating funds from both."[89] However, critics suggest the GCF, which generally promotes itself as a multilateral organization, functions in ways more like a multinational corporation that continues outdated development practices which leverage the power of global capital to provide communities what they are told they require, rather than

[88] Green Climate Fund, "Governing Instrument for the Green Climate Fund," 2012, 2.

[89] Sarah Bracking, "The Anti-Politics of Climate Finance: The Creation and Performativity of the Green Climate Fund: The Anti-Politics of Climate Finance," *Antipode* 47, no. 2 (2015): 285.

empowering them to engender ownership of their own development. Chris Methmann builds on this critique arguing that the way the GCF and similar emergent organizations operate opens political avenues that enable organizations and the countries that fund them—and thus to varying degrees influence their actions—to "claim to be in favour of climate protection and stick to business as usual at the same time."[90] This de-democratization of aid can lead to what Jacqueline Best describes as technocratic exceptionalism, as these types of supranational entities "operate, on a day-to-day basis, in political spaces exempt from many of the norms of liberal democratic politics and yet have the power to define and constrain them."[91] While Best is writing more specifically about central bankers and border guards, here I argue that a similar set of processes is occurring as actors like the GCF gain increasing power and autonomy within the international realm while operating in arenas that are often characterized by a lack of state capacity.

Sarah Bracking among others[92] has criticized these types of global institutions that are overly technocratic and resultantly become increasingly undemocratic. This critique builds on the work of scholars such as Thomas Pogge that have outlined patterns that began emerging following the end of the Cold War and into the 2000s in the international and supranational governance architecture where contradictions of interests and imbalances of power often—and perhaps increasingly—pit the interests of the Global South and Global North against one another. While this can be seen in a multitude of examples within global environmental governance—for the GCF as a specific example—this is most evident with powerful countries in the Global North providing most of the funding, and resultantly holding a high degree of decision-making power. On the other side of the coin, countries within the Global South are predominantly those in need of capital and assistance and resultantly the traditional donor-recipient relationship remains largely unchanged. The GCF is staffed by increasingly technocratic experts who have overseen the development of complex

[90] Chris Paul Methmann, "'Climate Protection' as Empty Signifier: A Discourse Theoretical Perspective on Climate Mainstreaming in World Politics," *Millennium* 39, no. 2 (2010): 345.

[91] Jacqueline Best, "Technocratic Exceptionalism: Monetary Policy and the Fear of Democracy," *International Political Sociology* 12, no. 4 (December 1, 2018): 328.

[92] See for example: Bram Buscher, "Anti-Politics as Political Strategy: Neoliberalism and Transfrontier Conservation in Southern Africa," *Development and Change* 41, no. 1 (2010): 29–51; Aihwa Ong, *Neoliberalism as Exception Mutations in Citizenship and Sovereignty* (Durham [N.C: Duke University Press, 2006).

technical engagements which have made the process from application to partnership, programme implementation, and ongoing management continuously more arduous. This technocratization also demonstrates the disciplinary power with which large organizations staffed with experts can leverage global capital flows in order to discipline the actions of developing communities in diverse and dynamic settings. While this can be seen to be productive in some instances, this process has served to exclude many who would otherwise be optimal candidates for funding.[93] Bracking argues that

> this is reflected in the little reflection by CSOs [civil society organizations] on what constitutes 'expert technical authority', or on how the 'calculative entities' and 'framings' that experts produce constitute knowledge, despite the role of these actors and processes in producing 'anti-politics', that is, rendering the process technical and thereby limiting open political and democratic deliberation.[94]

Those who work for the GCF are suggested to be experts, and thus their knowledge to be treated as gospel and their authority to not be questioned. As the GCF chooses partners to implement projects this then filters down from the top, leaving little space for local nuance, Indigenous knowledge, or effective partnership with local communities. While the GCF has undeniably contributed to many demonstrable benefits in countries and communities across the Global South, it has also seemingly done so in part through a continuation of certain neocolonial practices that have long been widely criticized within development discourse—though not always reflected in practice.

One example of the politics of such arrangements is with the awarding of grants by the GCF for sustainable development initiatives in the Global South. The process to apply for and be awarded a grant from the GCF is highly complex and requires a great deal of technical expertise and collaboration between a number of parties. The GCF does not implement projects itself directly, but instead works through partnerships with accredited entities. Accredited entities are responsible for presenting funding applications to the GCF and then overseeing, supervising, managing, and

[93] Jim Igoe, "Consume, Connect, Conserve: Consumer Spectacle and the Technical Mediation of Neoliberal Conservation's Aesthetic of Redemption and Repair," *Human Geography* 6, no. 1 (March 1, 2013): 16–28.

[94] Bracking, "The Anti-Politics of Climate Finance," 288–89.

monitoring projects once they are approved. The process of becoming an accredited entity in itself is long, complex, and takes a great deal of time and expertise. Once accredited, these entities work in close consultation with national development agencies and/or sub-national entities based on the various climate needs of individual developing countries and communities.[95] For Namibia this meant predominantly working with the EIFN as the main agent for the state along with the MET and involvement from other ministries as well as sub-national government input. The EIFN is also one of the main interlocutors that has cultivated a number of global partnerships to develop the types of technical expertise needed to put forth a successful proposal for such a grant. In so doing they must work closely with the CBNRM programme which means NGOs involved in a given region, local conservancy councils, and tourism organizations operating in the area.

CBNRM offers a number of potential benefits to various groups but can also create a number of concerns—similar to other approaches to sustainable governance—specifically regarding who benefits. For garnering funding through organizations and mechanisms such as the GCF, CBNRM broadly, and conservancies specifically, serves as an important channel through which to apply for and receive development aid. This is a result of CBNRM having been developed within development discourse and having emerged within recent evolutions in global governance. As there are already a number of global players working closely with conservancies, they become seen as both an effective inroad and in ways the path of least resistance for targeted funds to realize demonstrable goals that can be pointed to as successes for local communities as well as for the GCF itself and hence its donors (these donors predominantly being the wealthiest and more powerful countries). This type of 'success' tends to beget success as, in a world of increasingly squeezed national coffers, states are hesitant to donate aid unless it is tied to specific outcomes.

While none of this is surprising or inherently problematic, it can lead to unintended consequences that create issues for recipient nations and their people. For the specific case here, it has resulted in CBNRM becoming a programme that is in some ways 'too big to fail.' What this means is that as Namibia has been deemed a global leader in this form of resource governance, it has created a positive feedback loop of sorts where the more

[95] Green Climate Fund, "Project Lifecycle," Text, Green Climate Fund (Green Climate Fund, March 18, 2020).

success they can point to the more funding, tourists, and support they can attract. This cycle has created a more or less stable revenue stream that has kept the programmes functioning, though it also prevents it from evolving in order to achieve community autonomy and empowerment. These ideals are difficult to achieve as communities are disciplined to achieve certain outcomes in specific ways in order to receive benefits, without affording them the leeway that would be necessary for effective ownership to be engendered. There is so much interest tied up in CBNRM in its current and most well-known form remaining as such and continuing to function smoothly, that difficult questions about policy changes, political tensions, and power imbalances rarely get effectively discussed and issues are allowed to fester.

While making the CBNRM programme more environmentally sustainable is an important and necessary endeavour, deciding how this is accomplished remains a contentious issue. Partnerships that have emerged have led to measurable positive benefits; however, they have also created blinders of sorts and what some argue has led to a virtual straight jacketing of development efforts on the ground. Lara Diez suggests issues occur as the 'C' in CBNRM has come to refer more to *conservancy* rather than *community* and in the process

> forgetting the complexities and many needs/challenges of communities. Once conservancies are installed, they become a calling card and an end goal in themselves which ignores so many major issues. But resources go into conservancy activities because it is the buzzword for funding and then that's all that gets thought of.[96]

Here Diez is referring to conservancies themselves becoming an end, rather than a means to other ends—in this case, sustainable development goals. As a result, over the past two decades communities across Namibia have become disciplined to develop conservancies of their own and become gazetted. In some cases, this has made sense and many conservancies have become financially self-sufficient, with many even generating reasonable profits. However, for other communities, becoming a conservancy makes little sense and can in fact have negative implications. Hazam builds on Diez's point suggesting that many communities have applied for

[96] Interview, Lara Diez, Nyae Nyae Development Foundation of Namibia, January 10, 2020.

conservancy status and the MET grants it to them in order to point to their ever-growing numbers of conservancies as well as increasing percentage of land protected under communal spaces, but that often there are no local resources to manage.

With much of Namibia consisting of the Namib desert, many of the newest conservancies consist of territories that boast little or no natural endowments to be conserved or managed, and as a result a bureaucratic layer of governance is installed with no means by which to support it economically and little reason or impetus for doing so.[97] Nonetheless, the success of some leads communities to seek to emulate that success and apply for gazettement which is generally approved when the legislative requirements are met. This leaves these conservancies in a sort of limbo where they have a new system of governance, but with no meaningful way to staff or organize it and deriving no measurable benefits from the programme. In order to keep many of these going, NGOs and the Namibian government provide some degree of support and thus they drain important and scarce resources that could be allocated elsewhere much more efficiently. Therefore, conservancies ought to be seen only as one potential means to an end, rather than an end in themselves, and in a sense as a spoke with which to connect the local to the wheel of global funding streams.

The Sorris Sorris solar installation funded largely by the GCF is one example of a targeted project with heavy initial investment realized through partnerships between the conservancy, the lodge owner, NACSO, the EIFN, the MET, and the GFC itself. An example of another activity that is taking place to simultaneously combat climate change and mitigate human-wildlife conflict through the CBNRM programme is with an NGO called Lightforce. Based in France, this small NGO has grown over the years, and it provides opportunities for individuals who want to volunteer and get involved in fighting climate change and contributing to sustainable development in the Global South.[98] As part of its broader efforts, Lightforce sent a team to Namibia for three weeks between January and February 2020 in order to head to certain locations that had been determined would most benefit from small solar light installations being set up

[97] Interview, John Hazam, Senior Technical Advisor: Namibia Nature Foundation, January 23, 2020.

[98] More information on Lightforce can be found on its website at the following link: https://www.lightforceproject.com/

around livestock enclosures.[99] These decisions were made through collective deliberation between Lightforce, NACSO, the Ministry of Environment and Tourism, and through consultation with individual conservancy councils and members. As assemblages are ever changing, in this instance Lightforce thus becomes a part of Namibia's assemblage and imbricated in the governance of the CBNRM programme as well as its outcomes.

Lightforce is an individual component on its own but simultaneously takes on a new role in this time and space which subsequently has cascading impacts on all other components of the assemblage contributing to another evolution in response to climate change. As discussed previously, one of the major threats to CBNRM is human-wildlife conflict, as climate change has been pushing predators towards attacking livestock which in turn has led to farmers retaliating and killing wildlife.[100] Research has shown these solar light installations to be effective in decreasing the likelihood of predator attacks on livestock. They keep enclosures lit up throughout the night and can be set to flash the lights at various rhythms in order to mimic the presence of a human—suggesting danger to potential predators.[101] Thus, Lightforce members travel to Windhoek, procure as many of the materials as they can locally to support the domestic economy, while importing whatever they must, and from there head to rural areas in order to begin their work. While the activities of Lightforce are much more straightforward and standard NGO activities than the complex and costly dealings involved in the solar plant at Sorris Sorris Lodge, the novelty comes from the fact that the efforts of many local and global actors have shifted to countering the impacts of climate change while simultaneously treating issues exacerbated by it such as an increase in human-wildlife conflict.

Lightforce could have conducted their operations in the absence of conservancies; however, conservancies serve as legal governance entities and key actors that are empowered through denser connectivities to both Windhoek and the global community than rural communities traditionally had been prior to CBNRM. This served as an effective system for organization and communication by which Lightforce could work with other

[99] Interview, Team lead, NGO Team from France, February 3, 2020.

[100] Interview, Uatirohange Tjiuoro, Ministry of Environment and Tourism: Directorate of National Parks and Wildlife, January 20, 2020.

[101] Interview, Team lead, NGO Team from France, February 3, 2020.

NGOs operating on the ground as well as leverage the local knowledge and experience of conservancy councils in order to efficiently determine where the light installations were most needed and would provide the most benefit. In this sense as Kalvelage et al. argue, "conservation as a global strategy of environmental governance not only established new territories by commodifying landscape and wildlife, but also brings about new processes of territorialization."[102] This territorialization occurs by ordering and connecting these rural areas that were previously seen to be largely disconnected from the global, and instead codifies the territory and creates governing apparatuses in the image of Western Weberian states which are seen as more effective vehicles for developing partnerships for various purposes. The conservancies in this case are seen as pivotal in working with farmers to develop and improve the technology to ensure it realizes its goals of warding off predators and protecting important livestock populations.

While on a much smaller scale than the Sorris Sorris Lodge solar plant, the work of Lightforce has been positively received by conservancy members who see these activities as being a newfound benefit of the conservancy programme and a welcome green investment in their livelihoods and well-being. One farmer from just outside of the town of Bersig (Torra Conservancy), approximately 150 km Northwest of Sorris Sorris, reported that members used to

> get good benefits from the CBNRM programme except now with the drought we no longer get meat and also get less water. I grow some vegetables but it's very hard without enough water—same for livestock. The solar installations by the French team give us renewed hope and remind us of the good conservancies can bring and help save the few livestock I have left.[103]

While I was met with similar sentiments from a number of interview respondents, there remains the risk that so many rural Namibians are increasingly relying on the will of international actors, leaving them vulnerable to a loss of interest or to the next shift in approaches to development that might see efforts turned elsewhere. As a result, it is integral that conservancies coalesce behind the type of longer-term investments that were mentioned above by David and that they educate their

[102] Kalvelage et al., "Territorialising Conservation," 284.
[103] Interview, Farmer 1, Torra Conservancy, February 2, 2020.

children—especially girls—in order to develop the types of internal capacities needed to develop ownership over future development initiatives.

As climate change has continued impacting the assemblage, the need for these initiatives has become stark and the programme has thus embraced these partners who have in turn welcomed the programme which has proven useful for realizing both local and global goals. There remains a number of important questions however, regarding how resources are allocated, who chooses the types of projects that will be funded, and whether winners are being picked based on local needs and potential results, or they are being chosen as a result of their connections and/or what will most benefit the international community.

These partnerships do not simply fall into place haphazardly, but instead are the result of a complex set of decisions made by a multitude of actors whose efforts have coalesced. This can be beneficial and productive, but it can also lead to institutional and bureaucratic competition for scarce resources. It is for this reason that it is imperative to analyse the kinds of socio-ecological processes that were outlined in the previous section, and which are threatening the CBNRM programme. These processes are integral to understanding shifts in the components that make up the CBNRM governance assemblage.[104] While CBNRM is intended to empower communities, Duffy demonstrates that the monetization of resources can lead to resource abundance rather than scarcity leading to conflicts and can actually make stronger, not weaker, states.[105] This is seen with a concerted set of state actions being taken in the years following independence, with the NDP1 in 1995 that led to the creation of the EIFN, and subsequently the Nature Conservation Amendment Act being passed in 1996. This coalescence has occurred not by some miraculous feat of consensus in global politics, but instead is a result of endless negotiations and careful manoeuvring that goes into these types of highly political governance arrangements.

Novel Outlooks and Opportunities

As Chap. 2 outlined, the outlook on CBNRM has grown less optimistic over the years. Within the country itself, the programme remains popular and while many people acknowledge that it is far from perfect, the

[104] Katz-Rosene and Paterson, Thinking Ecologically about the Global Political Economy.
[105] Duffy, "The Potential and Pitfalls of Global Environmental Governance," 100.

majority of those I spoke to reported that their lives are better as a result of this form of resource governance.[106] It is not, however, without its critics whose voices have been strong in opposing the conservancy programme. CBNRM is far reaching and has become politically and economically engrained in the country. Despite this, its foundations are not as solid as some might desire and this type of political opposition can lead to cracks that can, and in ways have, eroded this foundation.[107] Increasingly however, as the drought has stretched on year after year people have realized that many of the challenges they are being faced with are a result of the effects of climate change which is presenting them with a new normal that they must adapt to. This has led to shifting outlooks on the programme that have begun seeing it in a newfound light.

While many respondents from conservancies acknowledge the hardships faced and the fact that things are not as easy as they once were, they similarly suggest that what they still have is largely thanks to economic opportunities and systems developed from tourism incomes as demonstrated in Fig. 6.2.[108] During an interview one respondent asserted that "conservation is very important; it is all we have left until we get rain. Without it we would have nothing, so we need to conserve for the future."[109] This was a sentiment that was shared by many throughout the dry fragile landscapes of the conservancies. For many, any sort of farming has become virtually impossible. Where some forms of agriculture had once been manageable, what they have left has been dying. Where sizeable herds of livestock once thrived, tens of thousands have died off as the land's carrying capacity dwindles by the year.[110] For rural populations who were never wealthy to begin with, it is difficult to communicate how hard it is to feed communities that have been experiencing population growth on land that is simultaneously able to produce ever-less. While the new partnerships that have been forming and the activities that have been developing in response to the challenges of climate change are not necessarily providing solutions to these complex problems faced by communities, they are providing opportunities that are helping to give a new

[106] Focus Group 2, Anabeb Conservancy, February 7, 2020.

[107] Interview, Lara Diez, Nyae Nyae Development Foundation of Namibia, January 10, 2020.

[108] Focus group 3, Anabeb Conservancy, February 7, 2020.

[109] Focus group 2, Anabeb Conservancy, February 7, 2020.

[110] Interview, Benedict Libanda, CEO, Environmental Investment fund of Namibia, January 29, 2020.

outlook on conservation in the country. Many Namibians have long believed that conservation is key to a sustainable future, but the drought has made it clearer than ever. They realize that what is lost today might not be able to be recovered tomorrow and collectively they are deliberately working with the assemblage of actors in order to invest in their future.

One of the keys to understand and that bears repeating again is just how vast Namibia is and how sparse its populations are. As a result of this, much of its populations live in extremely rural landscapes which can make any type of development extremely challenging. This chapter, however, has outlined some ways in which novel partnerships are leading to activities that are contributing to realizing certain sustainable development goals. As mentioned from the outset of this chapter, climate change as well as other challenges have made CBNRM based solely on wildlife management and ecotourism a precarious industry subject to volatility. Over the years however, conservancies have evolved beyond simply managing wildlife and attracting tourism and have become local-level governance entities that connect rural Namibian communities more intimately to the broader international political economy. Conservancies each have councils which bring organization and a platform for local communities to have a voice. In this sense communities are also able to exercise productive power based on both their geographic and historical relationship to wildlife which the paradigm of community-based conservation sees as central to effective development. While the level of organization varies across conservancies, each is required to write a constitution and present a resource-sharing plan to be gazetted. Thus, even in the most nascent, smallest conservancies whose governance capacities are least well-developed, there are mechanisms for democratic decision making and a platform for members to be heard. It is through this voice that resources are able to be directed to community benefits and efforts combined to enact change and move in necessary directions to ensure a sustainable future in light of whatever challenges might come their way.

While it is certain that benefits do not always materialize in an equitable and inclusive manner, the democratic mechanisms for resource-sharing plans as outlined in Chap. 4 do provide at least some means of transparency and accountability to conservancy members. CBNRM has in some ways quite literally put many of these places on the map. Previously, it would have been impossible for outsiders to find many of these communities through any of the online tools we have all come to rely on. However, now they have lodges and conservancy offices, and the local beauty and rich culture are increasingly accessible to global markets which can benefit local economies, but also have a disciplinary function as conservancies seek to fit the moulds of international capitalistic modes of ordering. While this increasing connectivity brings its own set of complex challenges and concerns, it is also enabling the types of partnerships outlined in this chapter. These partnerships can encourage the joining of Indigenous knowledge, with modern science, in order to develop sustainable solutions that are enabling people to continue to reside in their homelands, even as climate change threatens their traditional livelihoods. Eric Xaneb argues that

> Indigenous knowledge is better than scientific knowledge. Indigenous knowledge is human knowledge and human knowledge created scientific knowledge. Scientific knowledge should be brought in to supplement Indigenous knowledge, not replace it—especially in these harsh, distant climates that are each so unique.[111]

While scientific and Indigenous knowledge ought not to be pitted against one another in any sort of competition, nor are their joining a simple or always fruitful process, acknowledging the benefits both can bring and the foundations upon which they have developed are at the very least an important starting point from which to seek to develop effective solutions. Each form of knowledge is also highly politicized and is afforded particular forms of power. Siba Grovogui elaborates a particularly salient understanding of this discussion in his "revaluation of 'international knowledge' in light of recent scholarship in the fields of hermeneutics, ethnography, and historiography regarding the 'non-West' past, and the present of international society."[112] He like others does not see international knowledge as inherently problematic, but does problematize the way with which it is so often thrust upon the non-Western world to be taken as gospel. This form of knowledge comes with the forms of productive power I am outlining here and those on the ground who are intended to be recipients of such knowledge are incentivized to accept it as such by the benefits that often come with doing so—predominantly in the form of development resources. This is specifically the type of power imbalances that I am pointing to with this book and which must be acknowledged in order to realize improved developmental outcomes.

[111] Interview, Eric Xaneb, Chairman, Tsiseb Conservancy, January 27, 2020.

[112] Grovogui, Beyond Eurocentrism and Anarchy, 13.

While Western scientific knowledge has long been dominant within global environmental governance, the shift to community-based conservation has added the necessity of local Indigenous knowledge and practices. This has enabled local populations to exercise power as managers of valuable wildlife populations seen as a global good which inserts communities as key actors in the assemblage. While this does not erase power imbalances, it ensures that their voices remain central to development puzzles to some degree. Eric raises an important point that is in many ways central to the ethos of CBNRM. Part of the importance of the community empowerment pillar that puts local residents at the centre of development initiatives is intended to move away from the failures of cookie-cutter development initiatives that the international community pursued for decades. CBNRM is intended to enable global actors to partner with local communities in order to provide training and share technical knowledge but only to supplement their own existing approaches. CBNRM serves to erase some of these divides however, and analysing it as a policy assemblage enables us to move away from these distinctions and levels in order to understand the way actors operate at various levels and within diverse realms simultaneously. In this sense, communities which are so often understood as 'local' and divided from the global are indeed both at once as they meld their knowledge with other global partners and the two pressure one another into hybrid outcomes. Indigenous knowledge has emerged from centuries, and in many cases, millennia of experience with the local geographical, social, political, and historical realities which are all key to meaningful and successful development initiatives. Hassan Kaya and Yonah Seleti argue that many approaches to education in Africa are "still too academic and distant from the developmental challenges of African local communities."[113] There are endless important lessons that have been learned about conservation and sustainable ecosystem approaches that have been learned in Africa since the dawn of mankind and it becomes necessary to leverage these approaches and triangulate them against lessons learned from Western science and research to create positive outcomes. While marrying the two sounds good in theory, in practice we know this is both difficult to accomplish and leads to tensions and power imbalances. Nonetheless this joining of the local and global,

[113] Hassan O Kaya and Yonah N Seleti, "African Indigenous Knowledge Systems and Relevance of Higher Education in South Africa," *The International Education Journal: Comparative Perspectives* 12, no. 1 (2013): 30.

the scientific and Indigenous, are important pieces to every development puzzle and global assemblages provide the analytical tools required for understanding this complexity.

Conservancies have spread like a web of sub-national governance across the country in a way that did not previously exist. There had already been local traditional authorities and there are regional governments, but conservancies have developed somewhere in between these two and have served as an important inroad to make development initiatives increasingly sustainable. CBNRM has created a foundation from which the tourism industry has been able to proliferate. As the tourism industry has experienced and acknowledged the impacts of climate change on itself as well as its impacts on climate change, the assemblage of actors involved have begun working to evolve in response, in order to mitigate and adapt to these challenges. This has provided an effective inroad for renewed global partnerships that have provided resources for a number of development initiatives that are creating real and demonstrable benefits on the ground. Despite these opportunities, however, it is clear that these development initiatives do not serve to empower communities in the way proponents initially espoused or that some of the early literature suggested.

Conclusions

Eric suggests that "it is tough for people to not feel like the programme is always progressing, similar to the demand for eternal growth everywhere."[114] Many have questioned the notion of limitless growth on which the global capitalistic economy is based and there must certainly be limits to the growth of CBNRM in Namibia.[115] However, this does not mean that the programme is failing or doomed, but rather that it will continue to experience challenges that the actors that compose it must adapt to. Similarly, Maxi Pia Louis suggests that "drought is like any other financial type of crisis when people blame existing systems; in this case the conservancies and they say they don't work."[116] While drought might not be similar to every other financial crisis, her argument that when people

[114] Interview, Eric Xaneb, Chairman, Tsiseb Conservancy, January 27, 2020.

[115] Donella Meadows, Jorgen Randers, and Dennis Meadows, *Limits to Growth: The 30-Year Update* (Chelsea Green Publishing, 2004).

[116] Interview, Maxi Pia Louis, Director, Namibia Association of CBNRM Support Organizations, January 14, 2020.

struggle financially they often point to the system of governance which oversaw that financial hardship is a sound one. Economic struggles can be the result of poor governance, but can also occur when good governance is present but exogenous factors lead to economic downturns or prevent development. For the case of Namibia, climate change appears to be compounding every existing challenge to the CBNRM programme. While this might not be directly linked to every governance challenge, various existing issues have all been exacerbated by environmental changes. However, as was pointed out by several of my interview respondents it has also become an important learning experience and a necessary "kick in the ass" as suggested by the director of IRDNC at a full day workshop on impacts of climate change on the CBNRM programme.

This chapter has outlined many of the challenges faced by the CBNRM programme and the way these are all exacerbated by the impacts of climate change. The chapter, however, also turned from these challenges to analyse some of the opportunities that have been presented to Namibia's CBNRM programme in the face of a changing climate and its many impacts. As the international community has increasingly turned its efforts towards fighting global climate change, the actors involved in CBNRM have similarly been shifting much of their focus to greening their ecotourism activities. This has resulted in a number of opportunities for local, national, and global actors to develop novel partnerships in order to realize projects that contribute to the realization of sustainable development goals. These partnerships have produced fruitful outcomes such as the solar power plant at Sorris Sorris Lodge, and human-wildlife conflict mitigating solar lighting installed by Lightforce. One of the most interesting impacts climate change has been having is that as its effects provide inroads for new actors to join the assemblage and the broader CBNRM programme evolves, it is leading to productive power relations that are contributing to an increasing prevalence of the language of climate change which is simultaneously beginning to flow through their every action. While I did not go looking for climate change, it quickly found me and it has become a central theme that is inescapable in Namibia in general, and within CBNRM in particular. This has led existing actors to open up to novel partnerships in order to procure resources for initiatives to re-green ecotourism activities in order to remain competitive to continue attracting global capital. However, the disciplinary ethos of climate change has also led to a continuation, and in some ways perhaps an exacerbation of power imbalances that have important effects on the degree to which communities benefit from CBNRM which the next chapter will now turn to.

CHAPTER 6

Assembled Governance from the Local to the Global Community

In the penultimate chapter, I discuss the ways in which CBNRM is able to adapt to a variety of forces and who benefits as a result of these evolutions. This chapter unpacks the actors that are key in driving the evolutions which are integral to understanding why, how, and in reaction to what the programmes adapt. Part of this adaptation is importantly with the make up of the programme itself in the sense of who holds positions of power and how these positions relate to questions of gender and race. In order to understand this, we must analyse power imbalances that exist between a variety of actors and specifically between local communities and more powerful global actors who are often able to realize their interests. These shifting fashions that emerge within the international community end up having substantial impacts on the ground regarding what types of activities are favoured and hence funded. It becomes clear that going local is not sufficient in itself to prevent corruption and as a result, politics and power relations must be analysed from both the global and local levels.

> *CBNRM is actually more about community organizing than anything. Not specifically about money or wildlife but empowering the communities to make long-term development for themselves.*[1]

[1] Interview, Uatirohange Tjiuoro, Ministry of Environment and Tourism: Directorate of National Parks and Wildlife, January 20, 2020.

A. Heffernan, *The Global Politics of Local Conservation*, Environmental Politics and Theory,
https://doi.org/10.1007/978-3-031-24177-2_6

More than anything the threat of climate change has resulted in a shift to make the CBNRM programme greener in all its various facets. These efforts are important as the type of tourism that takes place has not traditionally been sufficiently eco-friendly to be considered sustainable at any kind of scale. While efforts to reduce the negative environmental impacts of the programme have been successful, they also call into question the reliance on development funding to improve the stay of wealthy tourists in Southern Africa, when nearby populations continue to struggle for subsistence in many cases.

As with any policy, CBNRM is not without its controversy nor without its critics. For a long time many Namibians were sceptical of this form of resource governance viewing it as simply the newest version of neocolonialism and as a way for foreign interests to control resources and subsequently local communities.[2] This scepticism was understandable given the violent and oppressive histories of colonialism and Apartheid and the country's long and bloody struggle for independence.[3] Over time as the programmes thrived, these attitudes shifted and CBNRM became favourably viewed by many Namibians which was evident throughout the interview process. Even those who admitted they were once sceptics now reported they believe CBNRM to be largely beneficial for the country.[4] There were also a number of interviewees who voiced some discontent with the way conservancy affairs were being managed but remained positive about the programme as a whole.[5] As the pressures of climate change have diminished many of the measurable returns from the programme, however, there has been a renewed uncertainty about CBNRM.

John Hazam, the Senior Technical Advisor of the Namibia Nature Foundation, suggested that this starts at the top as conservancy leaders were paid by NGOs early on but since then, as conservancies have been pushed to become financially self-sufficient, these salaries have had to decrease a great deal. Initially NGOs might have paid conservancy managers N$2400 per month (roughly US$240) but in many cases that might have been an entire conservancy budget. As NGO funding has tightened

[2] Interview, Dr. Margaret Jacobson, co-founder IRDNC, co-director, trustee, mentor, Conservancy Safaris Namibia, Swakopmund, February 17, 2020; Interview, Brian Jones, Environment and Development Consultant, Windhoek, February 20, 2020.

[3] Interview, Pohamba Shifeta, Minister of Environment and Tourism, January 29, 2020.

[4] Focus Group 2, Anabeb Conservancy, February 7, 2020.

[5] Interview, Barnabas, tour guide, Petrified Forest, Uisbasen/Twyfelfontein, January 31, 2020; Interview, Village respondent 1, Uisbasen/Twyfelfontein, January 30, 2020.

and conservancy manager and staff salaries have had to shift to be largely paid through tourism incomes, many have had to take sizeable pay cuts of up to 75 per cent. This has been one of the major challenges facing conservancies, as they are intended to emerge with the direct technical and financial assistance of donors and NGOs but are supposed to become entirely self-sufficient and eventually profit making. While approximately 25 per cent of conservancies have made this transition successfully, it is a difficult goal to realize, and the process remains complex and contentious especially as incomes have levelled off in recent years as shown in Fig. 6.3, despite growing populations and challenges to deal with.[6] People generally are not willing to continue doing the same work for less pay, Hazam suggests, especially in a world where the cost of living continues to rise.[7] Inflation has been a growing issue in Namibia—like with much of the rest of the world—and its causes are a combination of the growth in Namibia's money supply, aggregate demand, and the rising cost of food and fuel as well as supply chain issues which are exacerbated by relying on pass-through of South Africa for most of its goods.[8] All of these components are exacerbated by the large tourism sector, as wealthy tourists provide a steady flow of foreign capital, and many enterprises raise prices in order to capitalize on this. This is something I experienced first-hand as the majority of the food, and especially the fuel, that I purchased throughout my fieldwork were more costly than they would have been in many countries in the West. As a result, every purchase I made left me wondering how average Namibians are able to participate in the economy when the country's GDP per capita was approximately US$4100 in 2021, while also knowing that this is far above the income of the preponderance of rural Namibians due to the extremely high rates of inequality. While the majority of Namibians living on the subsistence or part-subsistence economy would not have eaten the same food as I did, these prices would make the cost of living a significant challenge for many (Fig. 6.1).

Beyond conservancy staff, members in general have been enjoying diminished benefits in recent years as well. It is important to note that

[6] MET/NACSO, "The State of Community Conservation in Namibia: Annual Report 2019: A Review of Communal Conservancies, Community Forests and Other CBNRM Initiatives."

[7] Interview, John Hazam, Senior Technical Advisor: Namibia Nature Foundation, January 23, 2020.

[8] Kaula Nhongo, "Namibia Hikes Key Rate by Most in 15 Years to Temper Inflation—BNN Bloomberg," BNN, June 15, 2022.

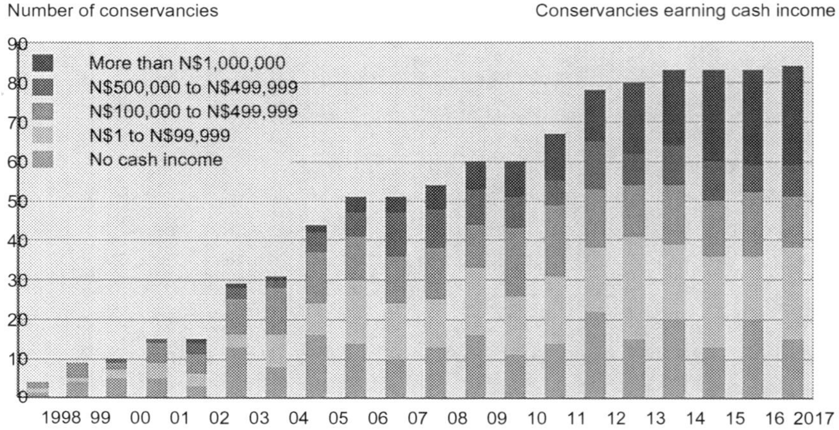

Fig. 6.1 Conservancy revenue 1998–2017 (MET/NACSO, "The State of Community Conservation in Namibia: Annual Report 2019: A Review of Communal Conservancies, Community Forests and Other CBNRM Initiatives.")

there are 92 conservancies across the country and the benefits and the way these are distributed vary greatly amongst them. Such differences aside, interview respondents regularly reported that many conservancies had seen these declines in recent years. One example from Torra Conservancy was that where conservancy members used to receive N$500 per year, recently it had decreased to N$50 or sometimes nothing at all.[9] Similarly, meat from game had been widely distributed but now was only being sent to the elderly or sick in the best cases. Hazam makes a key point in stating that when people's economic outlook declines, a desire for change of some sort often ensues. As a result, many are becoming increasingly unhappy with the programme. Blame has been put on the NGOs, the international community, the West, the Namibian government, and much has been made about local corruption at the conservancy level. This chapter will further unpack the complexities of the challenges facing the CBNRM programme in Namibia and how they are impacted by global power relations.

[9] Interview, Village respondent 4, Torra Conservancy, February 1, 2020.

Shifting Fads, Fashions, Funding, and Futures

Early successes for CBNRM witnessed wildlife populations rebound and across the country marked increases were being achieved in a number of species including lions, black rhinos, giraffes, zebras, elephants, springbok, gemsbok, and others.[10] Along with this came increased revenues to conservancies through trophy hunting and ecotourism activities. Beginning around 2011 however, these revenues have slowed and general benefits from the CBNRM programme began decreasing and have continued to do so in most cases ever since.[11] There are several factors that have contributed to this decrease in both wildlife populations and subsequent conservancy revenues. In the first few years of these declines, community residents largely blamed a mix of local governance apparatuses, the national government, and the international community. These sentiments are discussed by the research of the sceptics as outlined in Chap. 2. It is the sceptics who point to this layered governance as a concern, as government-issued quotas for trophy hunting began to decrease over these years. Challenges with this cumbersome, and at times problematic, layering of governance were shored up by data from several interview respondents who suggested their dealings with the Namibian government alone consist of the Ministry of Environment and Tourism; the Ministry of Agriculture, Water and Forestry; the Ministry of Fisheries and Marine Resources; and the Ministry of Land Reform as well as a number of regional governments. Beyond these layers of government bureaucracy are local governments, traditional authorities, and now conservancies.[12] This already cumbersome state bureaucracy must further govern in collaboration with a number of other non-state and supra-state actors for the effective functioning of CBNRM. This includes the network of NGOs, international multilateral organizations, the private sector, as well as new hybrid actors such as the Green Climate Fund. This has resulted in a complex governance

[10] Boudreaux, "A New Call of the Wild."

[11] Interview, Maxi Pia Louis, Director, Namibia Association of CBNRM Support Organizations, January 14, 2020; Interview, Marina Tavolaro, PhD candidate in Biological Sciences at University of Cape Town, January 15, 2020; Interview, Uatirohange. Tjiuoro, Ministry of Environment and Tourism: Directorate of National Parks and Wildlife, January 20, 2020.

[12] Interview, Lara Diez, Nyae Nyae Development Foundation of Namibia, January 10 2020; Interview, John Hazam, Senior Technical Advisor: Namibia Nature Foundation, January 23, 2020; Interview, Dr. Ekkehard Klingelhoeffer, Head of Department—Wildlife Management and Ecotourism, University of Namibia, January 13, 2020.

arrangement in which these various actors work together towards supposedly synergistic goals; however, there are often disagreements about what these goals should be as well as more importantly how they ought to be attained. Moreover, even when agreement is achieved there are always winners and losers as a result of choices made and actions taken. This leads to various entities jockeying for position to realize benefits, yet when it comes to responsibility for costs there is often a void.

As Chap. 4 discussed, conservancies remain dependent on the assistance of NGOs as well as funding that is funnelled through them by international organizations and foreign governments. Due to evolutions in global environmental governance, international financial flows have largely shifted from a core funding approach to capital being targeted at ever-more specific issues and geographic regions.[13] Catherine Corson argues,

> The 1980s and 1990s rise of neoliberalism laid the conditions for the formation of a dynamic alliance among representatives of the US Congress, the US Agency for International Development, environmental NGOs and the private sector around biodiversity conservation. In this alliance, idealized visions of NGOs as civil society and a countering force to corporations have underpinned their influence, despite their contemporary corporate partnerships. Furthermore, by focusing on *international* biodiversity conservation, the group has attracted a broad spectrum of political and corporate support to shape public policy and in the process create new spaces for capital expansion.[14]

With USAID being one of the central donors to Namibia's CBNRM programme from the outset, it was this coalescence of actors within the US that served as a major impetus for the programme's initial success. Over time, however, funding priorities have shifted, voices of opposition particularly in the US have grown louder, and this American environmental coalition as well as many like it elsewhere have seen an increase in roadblocks to them achieving their ends. Thomas Parks argues that this results in NGOs experiencing

> fluctuations in funding because of the constantly shifting priorities of their international donors. Without domestic sources… NGOs are forced to

[13] Toepler et al., "The Changing Space for NGOs."

[14] Catherine Corson, "Shifting Environmental Governance in a Neoliberal World: US AID for Conservation," *Antipode* 42, no. 3 (2010): 576.

> re-align their priorities with donor interests in order to compete for funding... The resulting asymmetry in donor-grantee relations often leads to a crisis of legitimacy and deteriorating effectiveness for the NGO.[15]

People, and subsequently organizations, want their money to go to saving a specific type of rhino in a particular area, or for protecting habitat for desert elephants in one region for example, rather than for broadly funding 'CBNRM in Namibia' as used to be the case.[16] Willie Boonzaaier reported, "You can see this with the increasingly activist nature of people in this age. They fly one specific flag and try to ensure it is held higher than all the rest."[17] One of the ways these politics have played out is with the issue of trophy hunting which provides a useful empirical example of the ways in which donor funding and the position of NGOs within it impact development efforts on the ground.

The changes in the fortunes of CBNRM can be partly attributed to climate change, as outlined in the previous chapter. However, they are also heavily influenced by shifts in global financial flows based on approaches to conservation, referent objects to be conserved, or specific geographic regions, countries, and locations that are in fashion in any given moment. One illustration of the way CBNRM is intimately connected to the global political economy is with the way in which these shifting fashions have had a substantial impact on trophy hunting. This has materialized as with an ever-sharper gaze over where and how funds are being allocated and how they are being deployed as well as with an ever-more watchful and activist civil society, organizations are increasingly hesitant to fund any activity that might be linked to trophy hunting which is worth unpacking in some depth here.[18]

At one point trophy hunting contributed to over half the revenue for the CBNRM programme across the country with some conservancies having received 80–90 per cent of their income from hunting-related

[15] Thomas Parks, "The Rise and Fall of Donor Funding for Advocacy NGOs: Understanding the Impact," *Development in Practice* 18, no. 2 (April 1, 2008): 213.

[16] Interview, Andrew Malherbe, COO, Save the Rhino Trust, Swakopmund, February 18, 2020.

[17] Interview, Willie Boonzaaier, Programme Director, Integrated Rural Development and Nature Conservation, January 24, 2020.

[18] Interview, Michael Knight, Transboundary Leader for KAZA, WWF, January 23, 2020; Interview, Richard Diggle, Business/CBNRM Specialist, WWF, January 24, 2020.

activities.[19] Thus, in lowering quotas, conservancies have been able to distribute less benefits in cash and in kind and are able to reinvest less into various social services in local communities.[20] Trophy hunting is one of the more complex issues related to CBNRM as well as one that effectively demonstrates the way the local is inherently linked to the global which impacts decisions on the ground. While, in general, quotas for the hunting of wildlife are issued by the state to conservancies based on an analysis of game numbers and approaches to sustainable harvesting, in practice the equation is not so simple nor is it apolitical. The Namibian Government engages the help of WWF-Norway to survey wildlife numbers and from this data calculates what they see as sustainable quotas. This creates discontent among some as many within local communities feel they are being unfairly robbed of their ability to obtain livelihoods as a result of distant government decrees that are supposedly based on numbers developed by European agencies. Thus, rather than the productive approach to power that CBNRM predominantly entails, hunting quotas demonstrate the state's persistent sovereign power which leads to sentiments of contradictions among some community members who felt they had been empowered through devolution. Rodgers Lubilo and Paul Hebinck unpack some of the ways many Namibians feel this amounts to national and global actors exercising their will at the community level, limiting the empowerment that may have been achieved.[21] Beyond this, however, as trophy hunting is a highly controversial activity internationally, the actions of conservancies at the local level can often be determined by global financial flows which are highly politicized. Prior to unpacking these complex global political economy issues, however, I will briefly discuss the issue of trophy hunting specifically as it pertains to CBNRM.

CBNRM generally consists of the sustainable management of wildlife in order to commoditize it for various tourism purposes. These generally fall under consumptive and non-consumptive use activities.[22] Non-consumptive

[19] Interview, Willie Boonzaaier, Programme Director, Integrated Rural Development and Nature Conservation, January 24, 2020.

[20] NACSO, "Resources & Publications: State of Community Conservation," NACSO, accessed May 7, 2020.

[21] Rodgers Lubilo and Paul Hebinck, "'Local Hunting' and Community-Based Natural Resource Management in Namibia: Contestations and Livelihoods," *Geoforum* 101 (2019): 62–75.

[22] Pascal Tremblay, "Wildlife Tourism Consumption: Consumptive or Non-consumptive?," *International Journal of Tourism Research* 3, no. 1 (2001): 81–86.

use activities are what most of us picture when we think of this type of resource management which involves protecting wildlife populations for ecotourists to engage in a sustainable form of travel that involves taking safari excursions to see animals, take photographs, and experience the rugged, rural Namibian terrain. The Nature Conservation Amendment Act suggests that "'non-consumptive use' means use not entailing the permanent removal of individual game, but use for recreational, educational, research, cultural, or aesthetic purposes."[23] On the other hand, the Act suggests that "'consumptive use' means the utilisation of individual game by its permanent removal, or removal of its parts, from within an area."[24] Consumptive use activities predominantly consist of various forms of hunting, where a different kind of tourist can pay substantial amounts of money to be issued a quota legally and be taken by a guide to track and shoot a specific wild animal. Until around 2010, for most conservancies trophy hunting made up the majority of their income and as a result it was one of the more popular activities.[25] While the vast majority of quotas issued are for various types of antelope rather than more charismatic megafauna such as the Big Five, "Hunting trips to shoot elephants (*Loxodonta Africana*) and buffaloes (*Syncerus caffer*) fetch high prices (USD 10,000–40,000) and account for most of the revenues generated from trophy hunting."[26] It is particularly the hunting of the Big Five which has led to trophy hunting becoming increasingly condemned by many within the global environmental movement. This condemnation in turn has led to donors targeting funding away from these activities as well as tourists choosing to spend their money at lodges in conservancies that focus on what are framed as, and hence what many accept as, less consumptive uses. While some might consider the killing of any animal for sport unjust, for Namibians it is a much more complex issue.

The shift in public opinion in the Global North in opposition to trophy hunting in Southern Africa has been exacerbated by extensive and disproportionate coverage of 'canned hunts' in which wildlife are bred in large, enclosed areas for the purposes of trophy hunting.[27] Specific events such as

[23] MET, "Nature Conservation Amendment Act," 3.

[24] MET, 2.

[25] Robin Naidoo et al., "Complementary benefits of tourism and hunting to communal conservancies in Namibia," *Conservation Biology* 30, no. 3 (2016): 628–38.

[26] Kalvelage et al., "Territorialising Conservation," 287.

[27] Heffernan, "Accounting for Climate Change in CBNRM: Reflections on Wildlife Conservation in Namibia."

the killing of 'Cecil the Lion' in Zimbabwe by an American dentist poured fuel on these movements which have accelerated as a result. While condemning the killing of Cecil is one thing, this increasing awareness and resultant outcry of many in the West has also had severe impacts that threaten to undermine the commodified forms of wildlife conservation which have become key to the sustainable development of many African communities.[28] Tremblay outlines the way in which trophy hunting has been deemed in the negative sense a consumptive use form of conservation as each person's 'use' of individual wildlife species directly decreases the quantity of that resource.[29] This is an understandable, clear, and straightforward argument that speaks especially to those outside Africa who generally experience the beauty of the continent's charismatic megafauna from their smart phones and condemn the thought of someone shooting these animals for sport. As a result, this framing of trophy hunting has been gradually gaining salience and NGOs are finding their funding increasingly targeted to specific measures in what results in various external actors, disciplining decision making within African communities which can be problematic for at least two specific reasons.[30]

First, when trophy hunting occurs it does so legally through the issuing of quotas that are developed based on scientific data in order to ensure populations are managed sustainably.[31] While it may seem counter-intuitive to kill any of the species of animal that might be endangered, sometimes there can be a problem animal or overpopulation in a given location and removing that animal or a small number of a herd from the area might be overall beneficial to the remaining population.[32] The essence of the CBNRM programme is for people and wildlife to coexist peacefully side by side and thus, if a particular animal is causing destruction to livestock

[28] Nicoli Nattrass, "Conservation and the Commodification of Wildlife in the Anthropocene: A Southern African History," *South African Historical Journal* 0, no. 0 (April 20, 2021): 1–22.

[29] Tremblay, "Wildlife Tourism Consumption."

[30] Brockington and Scholfield, "Expenditure by Conservation Nongovernmental Organizations in Sub-Saharan Africa."

[31] Stasja Koot, "The Limits of Economic Benefits: Adding Social Affordances to the Analysis of Trophy Hunting of the Khwe and Ju/'hoansi in Namibian Community-Based Natural Resource Management," *Society & Natural Resources* 32, no. 4 (April 3, 2019): 417–33.

[32] Jennifer Jacquet and Nicolas Delon, "The Values behind Calculating the Value of Trophy Hunting: The Values Behind Trophy Hunting," *Conservation Biology* 30, no. 4 (August 2016): 910–11.

or posing a threat to human populations it may be better to remove it than to have people rebel and kill many of the animals. While there are many avenues that could be pursued for removing a problem animal or mitigating the issues it might be causing, issuing a permit for a trophy hunter to remove it offers a way to infuse money into both the national and local economies, as well as to provide meat for local communities. Furthermore, while rebellion and the killing of multiple animals are far from a certain outcome, these types of instances have been experienced in many cases which are seen in the literature and further emerged from my interviews.[33]

Second, while it can sometimes be necessary to remove a problem animal, doing so can also provide many benefits that can be put back into conservation programmes. If a wealthy hunter from a foreign country pays US$20,000 to shoot a lion, that money can be reinvested in the community and in conservation programmes aimed at protecting the rest of the lion population.[34] Stasja Koot demonstrates the way that revenue from trophy hunting is not always enjoyed equally among conservancy members as well as points to the many critiques that government siphons off many of the funds before they ever reach the conservancy. As a result, US$20,000 can seemingly quickly evaporate. However, revenue from trophy hunting, as with photographic tourism does find its way to conservancies, has benefitted government coffers, and has continued to directly fund a number of important conservation services and entities across the country.[35]

Mike Knight, WWF Transboundary Leader for Kaza, suggests that trophy hunting amounts to sacrificing the few for the benefit of the overall population.[36] Local conservancy leaders at times expressed similar opinions that equally outlined the complexity of the issues. If wildlife limit the ability of humans to create sustainable livelihoods, history has demonstrated that in many cases—though not all—humans will turn to poaching as a means of survival and programme success will be threatened. The chairperson of Uisbasen/Twyfelfontein Conservancy suggested,

[33] Lubilo and Hebinck, "'Local Hunting' and Community-Based Natural Resource Management in Namibia." Interview, Employee, Wereldsend research outpost, February 3, 2020. Interview, Andrew Malherbe, COO, Save the Rhino Trust, Swakopmund, February 18, 2020.

[34] Interview, Ruben Portas, Cheetah Research Project. Windhoek, January 14, 2020.

[35] Koot, "The Limits of Economic Benefits."

[36] Interview, Michael Knight, Transboundary Leader for KAZA, WWF, January 23, 2020.

> If it's an increase of elephants there is an increase of consumptions, damages, food consumption and problems then it's good to reduce the number for where there are residents. [Elephants] consume a lot of water, take whole trees. They damage environment. They are like kids who want to try everything.[37]

Uatirohange Tijuoro from the MET went so far as to claim that it is unfortunate, they cannot sell ivory.[38] He explained that any country with elephants will always end up with a certain amount of ivory through elephants dying for a variety of reasons and at times even through legal trophy hunting. To not take advantage of this economically is unfortunate he argues, despite understanding the need for the ban on the sale of certain animal products as a result of the international political economy of illicit animal products.

Lara Diez, Director of the Nyae Nyae Development Foundation of Namibia, links the need to limit the consumptive use of wildlife populations to a type of climate injustice as she says, "The North did all the polluting, but the South pays. The North also did a lot of poaching but now the South must save the animals."[39] While many NGO officials operating in the Global South would share this position which carries with it some merit, it is clear that neither the case of combating climate change nor wildlife conservation is this simple. Specifically with the case of conservation, Africans have been heavily implicated in the over-exploitation of their resources as well as degradation of their local environments and pitting South versus North in these complex issues is neither entirely accurate nor productive. Instead, it is imperative to understand the trade offs which must be made with challenging issues regarding wildlife conservation and sustainable development. In Namibia, hunting has always been a way of life with some forms of hunting being sustainable while others were not. Figure 6.3 also shows the profitability that can come with such forms of sustainable trophy hunting. Several of the interview respondents from this research pointed out that trophy hunting is no different from the quotas that are issued sustainably in Western countries to hunt for deer, moose,

[37] Interview, Adolph Thaniseb, Uisbased/Twyfelfontein, January 30, 2020.

[38] Interview, Uatirohange Tjiuoro, Ministry of Environment and Tourism: Directorate of National Parks and Wildlife, January 20, 2020.

[39] Interview, Lara Diez, Nyae Nyae Development Foundation of Namibia, January 10, 2020.

bear, or other wildlife that reside there.[40] While there remain debates about this form of hunting as well in the Global North, hunting remains carefully regulated and through the continuing issuance of quotas people continue to possess the legal right to hunt (Fig. 6.2).

On the other side of the coin is non-consumptive use conservation, generally consisting of various forms of photographic tourism. Photographic tourism consists of travelling to remote locations in order to experience African culture, its landscapes, and take pictures of its majestic wildlife and their habitat. Of course, this is not consumptive in the same way as trophy hunting, as hundreds of tourists can take photos of a lion without it being detracted from the resource pool. Knight, however, argues that everything is a consumptive use and that carbon footprint, GHG emissions, water and other resource use, waste produced, and infrastructure development must all be factored in when analysing various forms of conservation.[41] Tourism is an industry with a large footprint—both carbon and otherwise.[42] In factoring this footprint in, even the most eco-friendly version of tourism that sees people travel from the Global North to southern Africa, this 'non-consumptive' use form of conservation begins to appear quite consumptive in its own ways. To be sure, trophy hunters produce similar footprints as they too fly, drive, eat, and drink; however, an individual hunter can provide up to 50 times the income for a conservancy as a photographic tourist.[43] There is a limit to the number of permits that can be issued based on annual wildlife population censuses and hence this form of conservation has strict revenue limitations build into it based on wildlife populations. However, these limitations are not entirely different from those that exist with photographic tourism as the tourist lodges in Namibia each have limitations on beds and thus the number of tourists any conservancy can welcome at a given time.

To further unpack the argument of Knight, it is worth taking a deeper look at tourism and its relationship with resource governance and sustainable development. Since the early 1990s, tourism in general and ecotourism in particular have exploded as global industries. The United Nations World Tourism Organization (UNWTO) reports that 1.5 billion

[40] Interview, Christine, Manager, Anabeb Conservancy, February 7, 2020.

[41] Interview, Michael Knight, Transboundary Leader for KAZA, WWF, January 23, 2020.

[42] [23] Heffernan, "El discurso oficial de las autoridades marroquíes y sus aliados en la ecologización de sus políticas públicas."

[43] Jacquet and Delon, "The Values behind Calculating the Value of Trophy Hunting."

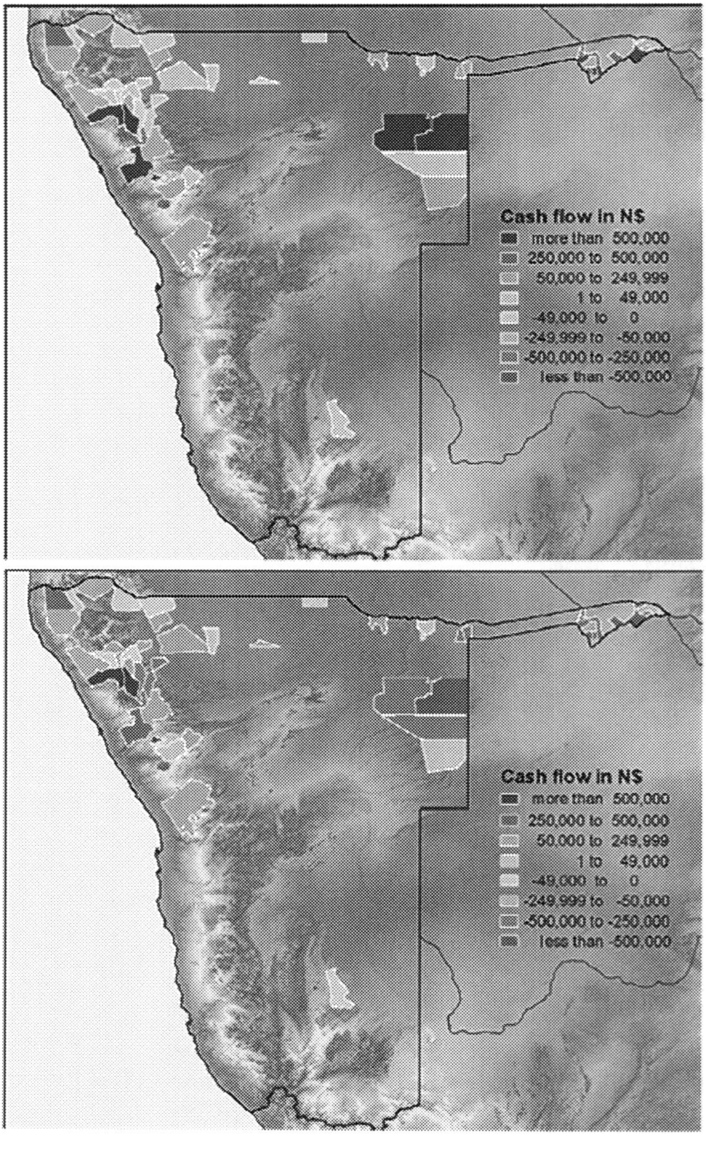

Fig. 6.2 Cash flows from hunting and photographic tourism (MET/NACSO, "The State of Community Conservation in Namibia: Annual Report 2019: A Review of Communal Conservancies, Community Forests and Other CBNRM Initiatives.")

international tourist arrivals were recorded in 2019. This marked a 4 per cent increase from 2018 and the same growth was predicted for 2020 and 2021 prior to the COVID-19 global pandemic that has rocked the industry. Pandemic aside, the UNWTO confirmed, "tourism is a leading and resilient economic sector, especially in view of current uncertainties."[44] The UNWTO's first comprehensive report on global tourism numbers and trends, the World Tourism Barometer, confirmed that 2019 marked the tenth consecutive year of growth, a year in which all regions saw a rise in international arrivals. Prior to the pandemic UNWTO Secretary-General Zurab Pololikashvili stated that "in these times of uncertainty and volatility, tourism remains a reliable economic sector." While in hindsight COVID-19 has demonstrated a previously unimagined volatility in the industry and undermined the validity of this statement, the industry is already rebounding and will likely continue to grow for years and likely decades to come. Already the live air traffic website FlightRadar24 saw June 15, 2022, set an all-time record with 232,483 flights that day, and they report we are on pace to continue breaking that record.[45]

Despite uncertain global economic perspectives, various geopolitical uncertainties, trade tensions, and social unrest, Pololikashvili affirmed the tourism "sector keeps outpacing the world economy and call[ed] upon [it] to not only grow but to grow better."[46] Growing better is key as the strength of tourism has often had it placed at the centre of development plans and with 2020 kicking off the decade of action to achieve the sustainable development goals by 2030, the time for better is now. This is the same deadline that the IPCC gave in 2018 when issuing a grave warning that the global community had a window of 12 years in which to cut GHG emissions by 45 per cent in order to prevent the rise in global temperature from exceeding the 1.5 degrees Celsius agreed to in the Paris Declaration.[47] Thus, while tourism can be an effective component of development, it must be ensured that it is done in as sustainable a fashion as possible, which is precisely what the evolving governance assemblage aims for CBNRM to become.

[44] UNWTO, "International Tourism Growth Continues to Outpace the Global Economy | UNWTO," 2020.

[45] Flightradar24, "Live Flight Tracker - Real-Time Flight Tracker Map," Flightradar24, 2022.

[46] UNWTO, "International Tourism Growth Continues to Outpace the Global Economy | UNWTO."

[47] IPCC, "Summary for Policymakers — Global Warming of 1.5 °C," 2018.

One of the concerns about CBNRM stems from all aspects of sustainability not being factored into various development activities. There is a rich body of literature that outlines the positive impacts tourism, and specifically ecotourism, can have for achieving sustainable development goals.[48] It is suggested by many from this camp that "ecotourism is promoted by governments of the North and South, international lending institutions, and private business, as an ideal development strategy that combines economic growth with environmental conservation."[49] Most of this is true. CBNRM can and has provided much in the way of economic development for a variety of countries across the Global South. Furthermore, CBNRM provides opportunities to empower local communities in order for them to realize development, as well as ownership which is a key component of sustainable development. However, for development to be sustainable, it must also be ecologically viable over the long term. In this, the impacts of ecotourism can become a quite complex phenomenon that requires careful analysis to understand the many costs associated with it as well as *who* benefits. CBNRM in Namibia has contributed in many ways to ecological conservation goals—specifically regarding the protection of wildlife and their habitat as has been demonstrated with wildlife population recovery as well as the improvements to land quality in areas that have been designated for wildlife mixed-use zones outlined in the previous chapter.[50] These are the many benefits often pointed to which have seen proponents argue that ecotourism activities are sustainable and synergistic with ecological goals. However, global environmental governance is a complex issue with manifold interacting actors which is also why the world has largely failed to act effectively to this point in mitigating environmental degradation and global climate change.

Even the most eco-friendly excursions that take place across Southern Africa tend to be GHG and resource intensive. Tourists arrive almost solely from the Global North, contributing a great deal in the way of aviation emissions. Once there, they generally rent large 4x4 vehicles or have one hired with a driver and spend a great deal of time traversing long distances in high-powered diesel trucks through the African savannah in

[48] Richardson, "Wildlife Utilization and Biodiversity Conservation in Namibia"; Jarkko Saarinen, *Sustainable Tourism in Southern Africa: Local Communities and Natural Resources in Transition* (Channel View Publications, 2009).

[49] Duffy, "Global Environmental Governance and the Challenge of Shadow States," 96.

[50] Coldrey and Turpie, "Climate Change Risk Assessment for Namibia's Communal Conservancies," 2020.

search of various ecological wonders. Tourists stay in hotels or upscale lodges, and research has widely demonstrated that people use more water and power, eat more, drink more, and generally live in excess when on vacation.[51] As such, it is imperative that scholars analyse the entirety of the costs associated with CBNRM including environmental impacts, in order to understand the effects CBNRM has on climate change, as well as the impacts the changing climate has on CBNRM. This is an exigent example of where analysing the assembled nature of CBNRM becomes necessary in order to understand the complex relationship between actors and their cascading and ever-changing impacts on one another as well as each other component of the governance arrangement which in turn impacts them. There is already a great deal of efforts on the ground being targeted towards improving the sustainability of ecotourism activities and these form a number of the important opportunities climate change is presenting CBNRM.

It is undeniable that the killing of wildlife is consumptive in a way that taking a picture is not. However, as argued by the director of NACSO, the ethos of CBNRM rests on the peaceful and prosperous coexistence of both humans and wildlife, and at times the killing of the few can be overall beneficial for the broader population of that species, as well as the ecological well-being of a given location.[52] While the question of overpopulation can be subjective, there are situations when it becomes clear to local people that issues are emerging based on a growing number of a given species such as lack of territory, veld food, water, and ongoing incursion into human settlements. In such a situation the local conservancy council or other forms of community leadership can apply to the MET for a permit for the removal of a particular animal or subsection of a larger population based on evidence which they must present. The MET will work with its partners to analyse the information from the application against data from wildlife censuses and conservancy audits and from there make evidence-based decisions. Allowing a limited number of people who pay significant amounts of money which can be reinvested into conservation activities in

[51] Saarinen, *Sustainable Tourism in Southern Africa.*

[52] Interview, Maxi Pia Louis, Director, Namibia Association of CBNRM Support Organizations, January 14, 2020.

the community can in some cases also have positive effects for the wildlife themselves.[53]

One example of such a situation was recounted to me by Ruben Portas, a researcher on the Cheetah Research Project who recounted that in 2019 an overpopulation of approximately 75 giraffes had developed in one area which was exceeding the geographic area's environmental carrying capacity. As a result, all of the giraffes as well as other wildlife and local human populations were being negatively impacted. The local community's traditional authority worked with the MET to procure five quotas which were in turn sold to private hunting outfits who then sold them to foreign customers for US$5000 per quota. These hunters travelled to Namibia spending a great deal more on accommodations and other tourist necessities bringing economic benefits to the local economies. Beyond this, once the giraffes had been killed, a truck was hired to bring local village men to the carcasses to butcher them and return a substantial amount of meat to local communities that benefited a great deal.[54] While to many the killing of such animals will seem unsavoury, this specific instance produced a number of benefits to local wildlife, people, and the broader CBNRM programme. The complexities of this type of case, however, can be difficult to communicate and often these types of nuances are lost in translation. This anecdote summarized in Fig. 6.3 also demonstrates the same logic that is employed for hunters in the West, whose activities are strictly regulated by ministries of natural resources and environment who issue quotas based on science and data targeted at ensuring healthy and sustainable wildlife populations that can peacefully coexist with nearby communities. By relying on global capital flows, however, the agency of communities and the NGOs with which they partner can be limited as productive power relations ply them to pursue certain forms of conservation more globally desirable, in order to continue to receive important resources and support.

What this discussion of trophy hunting can tell us is that NGOs and communities which are reliant on volatile and unstable funding streams are increasingly adaptive and responsive to the will of donors. In some ways this might be viewed in a positive light, as they could be seen as more nimble and adaptable to changes in the will of the global community. In other ways, however, it means that NGOs might be pulled from their roots and drawn towards interests that could have little to do with their

[53] Heffernan, "Accounting for Climate Change in CBNRM: Reflections on Wildlife Conservation in Namibia."

[54] Ruben Portas, Cheetah Research Project. Windhoek, January 14, 2020.

Fig. 6.3 Benefits of hunting (MET/NACSO, "The State of Community Conservation in Namibia: Annual Report 2019: A Review of Communal Conservancies, Community Forests and Other CBNRM Initiatives.")

core raison d'être as well as at times towards activities they might not even support. Kelly Krawczyk outlines an empirical case that demonstrates the way this has taken place in Liberia with civil society organizations (CSOs) upon which the country is highly dependent. As funding patterns ebb and flow, and as aid is increasingly targeted at the next global fashion, she argues CSOs—and resultantly much of the country—are faced with a great deal of volatility. This volatility makes it difficult for entrepreneurs to confidently engage in specific markets and as a result, certain industries in the country have experienced difficulties realizing success due to lack of consistency from a variety of angles.[55] Krawczyk outlines the important place donor funding can have on development, and the same can be seen with trophy hunting as a specific example as funding can ebb and flow based on a conservancy's decision to allow trophy hunting or not, as well as other choices made at the local level.

[55] Kelly Krawczyk, "The Relationship Between Liberian CSOs and International Donor Funding: Boon or Bane?," *VOLUNTAS: International Journal of Voluntary and Nonprofit Organizations* 29, no. 2 (April 1, 2018): 296–309.

I want to be clear that the point of this discussion of consumptive versus non-consumptive use of natural resources has not been to argue in favour of trophy hunting over photographic tourism, nor to support claims that it might in fact be less consumptive. Instead, the discussion is intended to outline the way global financial flows can have a direct impact on disciplining NGO activity and hence development practices on the ground. Furthermore, global financial flows are heavily impacted by shifts in global sentiments which results in productive power that determines what type of conservation in which specific region will be favoured in a given period. These 'fashions' can last a few years or even decades, but they are always changing and the lives of those in recipient nations are continuously impacted. What I am arguing with this section is that as new destinations, specific endangered species, and even the killing of a particular lion by an American dentist gain traction on the world stage, their importance can expand and see large swings where capital is targeted as well as development efforts and tourist dollars and interest. It is this ability of actors from all over the world, including those in their living rooms in the West, their governments, NGOs, international financial institutions, and a plethora of actors in the environmental movement to exercise productive forms of power and influence decision making and activities on the ground—though to different and varying degrees—that this discussion of trophy hunting sheds light on. What this means also is that conservation decisions should be made based on evidence and data, rather than politics and the next global fashion. While separating politics from policy does not happen in practice, acknowledging the politics involved in policy making, including the allocation of funding, is an important step in maximizing policy impact and minimizing the risks of unintended consequences. What is at stake here is power, knowledge, and the ability of competing actors to realize their interests, rather than finding the neutral truth.

As power is reorganized through CBNRM, rather than resulting in a subtraction of power from the state and an equal and opposite empowerment of communities, power becomes increasingly dispersed through a layering of governance across a network of actors who often have synergistic interests, while at other times they are disharmonious. What this results in is a complex situation where stagnating wildlife populations are leading to decreased quotas being issued, which are further gaining downward pressure from the international community. This is resulting in conservancies being hesitant to engage in trophy hunting activities even when quotas can be obtained. While it can be inferred that this is partially attributable to the CBNRM programme adapting to the changing climactic realities in Namibia as laid out in the previous chapter, some of the literature on global environmental governance suggests that this is also partly a result of these communities being intimately tied into the broader global power relations that develop based on the social interactions of countless actors across the international community.[56] It is also an illustration of democratic politics and the aim of proponents of CBNRM to prevent the tragedy of the commons. If each community had full sovereignty and power over their local resources, there would be the risk that some would extract them unsustainably which could lead to a depletion of wildlife as well as other important resources. While not everyone would agree, the environmental argument would suggest that this wildlife belongs to the planet and future generations and thus their importance goes well beyond the immediate community.

Conservation in general, and particularly in the context of development, is highly political. There is hard science that says we need to conserve now in order to protect our natural environments for the future, but there remains a great deal of debate about how this ought best to be accomplished. While at one time the consensus favoured fortress conservation, this paradigm eventually shifted to community-based activities that saw local populations integrate resource management into sustainable livelihoods.[57] As various community-based approaches have proliferated and eventually emerged as a virtual consensus within development discourse, subsequent questions have emerged about the degree to which communities should have autonomous decision-making power over their resources and land.[58] One key example of this is discussed above regarding the complexities of trophy hunting. While community empowerment is integral to CBNRM, devolution comes with its own risks and could result in a tragedy of the commons in which communities over-extract wildlife populations. As a result, a difficult balance must be achieved between local autonomy and national and global priorities. This question led to classifications of consumptive versus non-consumptive use activities, which is a

[56] Duffy, "Global Environmental Governance and the Challenge of Shadow States"; Duffy et al., "Why We Must Question the Militarisation of Conservation."

[57] Kathleen Galvin, Tyler Beeton, and Matthew Luizza, "African Community-Based Conservation: A Systematic Review of Social and Ecological Outcomes," *Ecology and Society* 23, no. 3 (September 11, 2018).

[58] Heffernan, "Accounting for Climate Change in CBNRM: Reflections on Wildlife Conservation in Namibia."

specific type of framing that serves to bring into view the politics of the issue and creates 'right' and 'wrong' approaches to conservation. Thus, a new consensus developed in the literature that was clear that when attempting to enact 'conservation' something 'consumptive' must surely be undesirable and counter-productive.[59] For many, this designation makes sense.[60] For with consumptive use forms of management such as trophy hunting, each animal killed limits the ability of others to travel to Southern Africa to enjoy that animal's beauty. As the academy remains largely Western-centric in its theories, institutions, and its scholars,[61] this framing largely dominates the literature, and a focus has been placed on photographic tourism while trophy hunting has been increasingly squeezed out over time.[62]

As discussed, CBNRM success has begun to wane over the past decade and the framing of consumptive versus non-consumptive conservation has led to questions about how and why this becomes such an important factor. Trophy hunting remains one of the leading causes pointed to, which is also often associated with poaching—despite these being two very different activities. However, trophy hunting has been demonstrated to be able to take place sustainably through limited quotas based on scientific research of wildlife populations.[63] This framing ignores African history and culture, as well as the real and present challenges that living with these often dangerous and destructive animals presents. Furthermore, as mentioned trophy hunting is often associated with poaching despite the fact that poaching remains illegal and increasing efforts have gained marked success in recent years to eliminate poaching entirely in some conservancies in Namibia.[64] Estimates vary, but most studies suggest only 1–2 per cent of key wildlife populations are killed by poaching every year in Namibia which does not pose great threat to most species.[65] While it

[59] Tremblay, "Wildlife Tourism Consumption."

[60] Richardson, "Wildlife Utilization and Biodiversity Conservation in Namibia."

[61] Arlene B. Tickner, "Core, Periphery and (Neo)Imperialist International Relations," *European Journal of International Relations* 19, no. 3 (2013): 627–46.

[62] Koot, "The Limits of Economic Benefits."

[63] Naidoo et al., "Complementary benefits of tourism and hunting to communal conservancies in Namibia."

[64] Lubilo and Hebinck, "'Local Hunting' and Community-Based Natural Resource Management in Namibia."

[65] Jones, "Policy Lessons from the Evolution of a Community-Based Approach to Wildlife Management, Kunene Region, Namibia," 1999.

would of course be preferable to eliminate any form of poaching entirely, these numbers demonstrate a vast improvement over years past and do not pose a great threat to wildlife populations. It is also important to analyse these numbers relative to if the programme was not in place, as it is likely these numbers would be much higher.[66] Nonetheless, there are those who frame the recent declines in CBNRM success as a result of continued poaching and over-hunting of wildlife populations which simply is not the case.[67]

Conclusions

While it is clear that CBNRM has adapted in response to the impacts of climate change in a number of ways which have been both positive and negative, evidence suggests that it is local communities which most often experience the greatest of the negative impacts. Economic benefits often fail to remain in the community, and it is simultaneously communities that most often experience unintended consequences in the form of environmental degradation, costs associated with wildlife conservation, and an undermining of their local-level governance capacities. These negative outcomes consist of certain wildlife populations that are locally less desirable, being favoured over others, autonomy in decision making over conservation activities being limited, economic benefits failing to materialize within the community to the degree expected, as well as a general deficiency in the level of empowerment enjoyed. These outcomes are a result of the power asymmetries that are built into local-global political dynamics which result in global actors exerting a disproportionate degree of productive power in relation to local actors and thus seeing their interests win out over others. Analysing this complex assemblage of actors as well as their competing interests from the position of a flattened ontology enables us to see the ways in which various issues are politicized and put on the agenda—such as trophy hunting and 'consumptive use' approaches to conservation. It also enables us to see the way 'non-consumptive use' approaches are depoliticized, and their ecological and climactic impacts

[66] Interview, Marina Tavolaro, PhD candidate in Biological Sciences at University of Cape Town, January 15, 2020.

[67] Ikanyeng Gaodirelwe, Moseki Ronald Motsholapheko, and Gaseitsiwe Smollie Masunga, "Community Perceptions of Wildlife Management Strategies and Subsistence Poaching in the Okavango Delta, Botswana," *Human Dimensions of Wildlife* 25, no. 3 (May 3, 2020): 232–49.

kept off the agenda. Each is pursued in order to benefit certain groups, with the predominant voices in the West and the global environmental movement generally winning out, and as a result trophy hunting—which is generally locally desirable—being squeezed out in favour of photographic tourism. As discussed in the chapter, this is not inherently problematic, however it raises difficult questions about the degree to which this form of global environmental governance is truly community-based, as well as who its benefactors are, which seem to be Western tourists as often as not.

CHAPTER 7

Development, Climate Change, and Global Assemblages

While it is clear is that climate change has presented a number of challenges to the CBNRM programme, it has also presented a number of opportunities which are being harnessed by certain actors which among other benefits is helping to diversify those holding positions of power. Climate change is challenging the viability of the programme itself while also creating avenues for new types of actors such as the Green Climate Fund who are afforded legitimacy and funding by the international community, but who do not function in the same way as traditional multilateral organizations. Thus, long after communities were supposed to have become self-sufficient, what we are seeing is a proliferation of actors involved, and a failure for communities to attain autonomous self-governance. As a result, CBNRM might best be described as International Community-Based Natural Resource Governance in which a complex assemblage governs resources at the local level. While in some ways, ground continues to be gained on this path, in others, this autonomy seems to be a distant goal that will remain unachieved.

> *Like democracy, CBNRM is both an idealistic theoretical goal and an imperfect practical struggle towards this goal.*[1]

[1] Child and Barnes, "The Conceptual Evolution and Practice of Community-Based Natural Resource Management in Southern Africa," 284.

A. Heffernan, *The Global Politics of Local Conservation*, Environmental Politics and Theory,
https://doi.org/10.1007/978-3-031-24177-2_7

Eric Xaneb's family have been subsistence farmers for generations but due to racist and oppressive land laws, mixed with a hot, dry climate, and unproductive soils, life had always been difficult for them. However, following independence there was renewed hope that Namibians could finally take ownership of their own governance and part of this would be realized through a redistribution of land and its resources to be shared communally through conservancies.[2] With this came the CBNRM programme and an energy and fervour developed along with the promise of sustainable development that would be locally owned and operated in partnership with a number of actors who would work towards congruous goals. Initially the programme experienced marked success as what would be recognized by most as economic development was achieved, wildlife populations increased, and some form of local-level empowerment was attained.[3] For Eric and many like him, this provided a relatively good paying, stable job as manager for Tsiseb Conservancy. He is honoured to have been elected to this position by his neighbours and fellow conservancy members, as well as grateful for having the opportunity to work towards conservation for the future. Regarding the future, he also holds out hope that his daughter might have prospects for a decent livelihood close to home. For years, many of Namibia's youth have been leaving rural areas such as the Erongo Region in search of employment outside of farming and they have been migrating to larger cities such as Windhoek, Swakopmund, or Walvis Bay in search of what they see as a better life and more modern forms of employment. Eric and many like him see CBNRM as a means of attaining sustainable development goals, ensuring natural resources will remain abundant for years to come, and that their natural environments will continue to remain ecologically stable.

For over two decades people like Eric benefited in a number of ways from CBNRM. However, the programme was not the panacea many of its supporters espoused and a number of challenges emerged over time. These issues were then exacerbated by the ongoing drought which has exposed many of the weaknesses of CBNRM while placing a great deal of strain on the programme itself. In short, the effects of the changing climate have resulted in less rain and hotter temperatures which are leading

[2] Brown and Bird, "Sustainable Natural Resource Management in Namibia."

[3] Jean-Gael E. Collomb et al., "Integrating Governance and Socioeconomic Indicators to Assess the Performance of Community-Based Natural Resources Management in Caprivi (Namibia)," *Environmental Conservation* 37, no. 3 (2010): 303–9.

to less veld food available for livestock as well as wildlife.[4] This means less prey for predators who are increasingly driven towards human settlements they would normally prefer to avoid and an increase in attacks on livestock that are already dying in large numbers due to a lack of water and food.[5] Less wildlife also means less tourism and thus as the programme itself is distributing fewer resources to conservancy members, it is simultaneously losing support from farmers who blame the wildlife for livestock losses.[6] For Eric, this creates a multitude of challenges. Eric and his family still grow what food they can and continue to practise pastoralism as a way of raising livestock. However, both of these activities have become increasingly challenging over the years, and as the drought has worn on, farming has become almost impossible throughout much of the Erongo Region as well as much of the rest of the country. As conservancy manager, challenges are also increased as he must govern a growing conservancy, with decreasing revenues which must simultaneously be stretched further in order to compensate for livestock losses, and continuously develop infrastructure to procure water in innovative ways.[7] The challenges go on and on and have been discussed at length in Chap. 5 but it bears re-iterating here the extent of the strain posed by climate change which many Namibians recognize as an existential threat.

While the challenges are almost limitless, and while no one would argue that climate change is providing any sort of net benefits to Namibia, it is and has been providing a number of opportunities which are important to analyse. It is relevant to note that opportunities in this context do not necessarily connote something intrinsically positive, but instead indicate how climate adaptation and necessary responses to climate change do provide avenues for some actors to adopt new strategies for profitable enterprises and survival. Namibia's version of CBNRM emerged almost solely around the conservation of wildlife for the purposes of ecotourism. Over time, Namibia has become host to 92 conservancies which cover over 20 per cent of its territory (166,045 square km) and over 15 per cent of its population (227,941).[8] Beyond that, almost every part of the economy

[4] Coldrey and Turpie, "Climate Change Risk Assessment for Namibia's Communal Conservancies," 2020.

[5] World Bank, "Overview," Text/HTML, World Bank, 2021.

[6] Interview, John Pallet, Director, Namibia Environment and Wildlife Society, Windhoek, February 21, 2020.

[7] Interview, Eric Xaneb, Chairman, Tsiseb Conservancy, January 27, 2020.

[8] "Registered Communal Conservancies," NACSO, 2019.

has become linked in one way or another with the programme, as CBNRM is so much more than the tourism industry but instead a political economic system of governance and organization that has spread out to encompass much of the country. The emergence of this system of governance was largely productive over time, but it also created weaknesses, as it became such an important pillar for the economy but was based almost solely around the management of a specific resource which has both long been threatened and which is highly mobile and often transnational in nature. The challenges of climate change have pushed the assemblage of actors involved in CBNRM to adapt in a number of important ways. Not only have they sought to evolve economically but in political ways as well. Stemming from these evolutions, this book has provided answers to the questions of how CBNRM has evolved to respond to the challenges of climate change and to what extent these changes benefit local people? Outlining the ways in which the programme is able to adapt to issues related to climate change is a key part of the response to the first research question which I will further unpack here prior to discussing how these evolutions occur, as well as to what degree communities benefit as a result.

Efforts at economic diversification have meant developing novel revenue streams through new kinds of tourism beyond photographic wildlife tourism and the more traditional trophy hunting incomes. This has included developing various types of tourism activities aimed at broader demographics. Part of these efforts at diversification have consisted of the emergence of a developing industry for tourists to learn more about and submerse themselves in the many Namibian cultures of both past and present. Living museums have emerged over the past decade which enable tourists to see how the many different peoples lived over centuries on the land now called Namibia. Increasingly diverse food and drink are offered as part of these experiences as well as crafts and wares that can be purchased at more and more stops along the extremely remote roads of the Namibian savannah as shown in Fig. 7.1.

These might sound like subtle changes but due to their nature have created livelihoods for more people than would previously have been possible as traditional CBNRM tourist activities create only a limited number of jobs working at lodges, safaris, and campsites. These emerging industries, however, have created the ability for many Namibians to derive some sort of income—albeit in many cases limited—as a result of the programme. Enabling more people to benefit from these programmes is key, as one of the binding constraints to programme success that has been

Fig. 7.1 The sale of artisanal crafts and wares (MET/NACSO, "The State of Community Conservation in Namibia: Annual Report 2019: A Review of Communal Conservancies, Community Forests and Other CBNRM Initiatives.")

pointed to in the existing literature is that while CBNRM has produced economic development opportunities, these remain limited to the few.[9]

Often, opportunities have been harnessed by those who were already most well connected in their communities, who had the highest levels of education, and importantly those with sufficient levels of English proficiency. As a result, those most marginalized in society tend to be precluded from benefiting in any meaningful way. For instance, Joni Seager, professor of Global Studies at Bentley University who works closely with practitioners on the ground in Namibia, reported that while these types of partnerships often result in benefits, these might be in the form of compensation for agricultural plots that are destroyed by wildlife. However, compensation only pertains to 'official plots' of over one hectare, and thus those less well off—specifically women—who tend to have plots much smaller than this, and are those who need the compensation most, receive none.[10] While these new opportunities have not lifted Namibia's marginalized populations from poverty, they are providing some with the ability to develop new skills and create meaningful livelihoods.

Beyond new forms of economic income, climate change has also pushed the more traditional side of CBNRM to evolve in order to become greener,

[9] Hoole, "Place-Power-Prognosis."

[10] Interview, Joni Seager, Professor of Global Studies, Bentley University, January 24, 2020.

which has resulted in increased business for those who do so successfully. A great deal of tourism cannot be said to be sustainable as a result of long flights, time spent in vehicles, resource consumption, and the creation of waste. However, ecotourism aims to minimize the footprints of tourists while investing incomes derived from them back into important conservation priorities. Thus, in order to cater to environmentally conscientious tourists, CBNRM tourism operators have collaborated with a number of partners in order to green the services they offer. One of the most effective ways they are doing so is with the greening of the lodges themselves. One of the more obvious examples of this is with the solar plant developed at Sorris Sorris Lodge that was discussed in Chap. 5, which saw the efforts of a large number of actors coalesce in order to build this source of renewable power. Beyond the solar plant, the lodge itself was designed and built with predominantly locally sourced materials and person power and was engineered in a way to consume as little power as possible. The lodge and its bunkhouses are high on a hill overlooking a vast plain which results in breathtaking views and an effective way to view wildlife, but also results in extremely hot sun beating down on the buildings for many hours of the day. Thus, the bunk houses themselves were constructed of a special kind of stone and made very thick, as well as partially sunken into the hilltop in order to leverage the natural landscape to minimize both heating and cooling demands of guests.

Other lodges have been taking the initiative to innovate in a similar way often at smaller scales but at their own costs. Okahirongo Elephant Lodge in Purros in the far Northwest of the country built their own solar plant which provides most of the power they need, and the rest is supplied by a backup generator. Similarly, Palmwag lodge which is owned and operated by the Gondwana Collective—a domestic company—has also been scaling up its solar generating capacities, though still relies heavily on their diesel generator as the lodge itself continues to expand and demand ever-more power. In recent years, the entire Palmwag compound has also been retrofitted with a new greywater collection system in order to ensure that water demands are met without tapping into valuable and scarce drinking water sources. As the greywater collection system demonstrated success, its technology and implementation were easily transferable throughout the conservancies and were quickly implemented, first at many of the tourist lodges and since then various versions have increasingly been installed throughout the country at both tourist and non-tourist operations, buildings, and dwellings alike. As it has proliferated, the systems have improved as so many have contributed various innovations to them, the country is now collecting more and more greywater for sustainable reuse. This form of drought proofing is both increasingly essential, and prior to supplementary income derived from the CBNRM programme, was not possible on any kind of broad basis as

> poor households are, by definition insecure, and in Namibia all rural households are vulnerable to drought, as rainfall is highly variable (particularly low and unreliable in the west). Therefore drought-proofing is an important way of reducing vulnerability. The cash income generated by tourism can be particularly useful for this for two reasons: first, timing: it does not slump when drought comes (although it is likely to slump according to market cycles rather than rainfall). Second: who earns it. Tourism can be one of the few sources of collective income, which can be usefully spent on helping members cope with drought.[11]

While these innovations might seem like an obvious response to drought, as well as transference of an emerging technology, the initial system would likely not have been developed without the CBNRM programme and without the pressures of climate change. Furthermore, it would not have continued to be improved or proliferated as quickly without the conservancies as well as the many partners they work with.

Beyond lodges, climate change has seen new opportunities for partnerships with actors who seek to support the conservation efforts of Namibians while also contributing to development and fighting climate change. An example of this is the NGO Lightforce that installs small solar lights at farms to deter predators at night.[12] The Head of Environmental Services of #Khoadi !Hoas Conservancy pointed specifically to these installations as having provided a measurable impact that has been well received by their conservancy members and which was making a noticeable difference in human-wildlife conflict.[13] While a seemingly small endeavour, this organization has made specific use of the governance network and the many partners involved in CBNRM in order to determine where these lights are most needed, how they would get to them, where the team would stay, and where the best places to source materials locally were. While climate

[11] LaFranchi and Ashley, "Livelihood Strategies of Rural Households in Caprivi," 21.

[12] Interview, Team lead, Lightforce, February 3, 2020.

[13] Interview, Albert Kroses, Head of Environmental Services, #Khoadi !Hoas Conservancy, February 10, 2020.

change has exacerbated issues of human-wildlife conflict, it has also pushed people to develop innovative solutions that have helped to improve a problem that has existed for centuries.

These efforts to increase the sustainability of ecotourism link directly back to the impact global politics have on the types of conservation activities that are favoured as elaborated by the discussion of trophy hunting in Chap. 6. As outlined, the purpose of this discussion is not to argue in favour of one or the other types of conservation. The contribution is instead to discuss the complexities of both trophy hunting and photographic tourism in order to analyse them as part of local and global political debates. While much of the world was rightly outraged at the killing of Cecil the Lion in Zimbabwe, that does not mean that all forms of hunting should be eliminated in Namibia, or across Southern Africa, or even that doing so would be beneficial for either human or wildlife populations. Instead, all forms of hunting should be carefully regulated based on scientific data of the sustainability of wildlife populations, carrying capacity of certain geographic regions, as well as the difficult balance that must be struck between people and animals living in close proximity and relying on the same healthy ecosystems. In this sense, I argue there is a place for limited trophy hunting in certain instances, as doing so can contribute to the overall health of wildlife populations while also creating revenue to be invested in conservation programmes and community benefits. While proponents of CBNRM report that these types of decisions continue to be made locally, and autonomously from outside interference, in employing the assemblage approach we instead see the ways in which global fashions can diffuse into local politics. This results in productive power that is exercised through the social relations that emerge from CBNRM and which globalize decision making based on the competing and at times simultaneously harmonious interests of heterogeneous actors. For this specific case, this means that Western donors, the environmental movement, and governments channel funding to programmes, conservancies, and conservation activities that favour non-consumptive-use forms of conservation and as a result inadvertently punish those who continue trophy hunting.

Similarly, regarding photographic tourism, my intention is not to suggest that this form of 'non-consumptive use' conservation ought to be eliminated or curtailed in any significant way. My aim is to point to the fact that it is consumptive in its own way, and that the negative environmental impacts of this form of tourism must be part of every ongoing discussion in order to minimize their impacts. Instead, what has predominantly occurred is those whose interests align with ecotourism have seen their voices win out, and questions of climate change and the consumptive nature of photographic tourism have remained off the agenda. Despite the consumptive nature of it, 'eco'tourism has been broadly deemed non-consumptive, and in the same breath has been depoliticized and is largely presented as an unquestioned good. Analysing these forms of conservation is key to unpacking ways in which the programme is evolving in response to climate change as well as demonstrating how actors within the assemblage are impacted and in ways produced by others and in turn how this influences outcomes. While discussion of climate change has been largely omitted from the literature on CBNRM, actors involved on the ground have long been coalescing in order to develop innovative solutions to lessen the impact of tourists on both fragile local environments and the broader challenges of global climate change.

Environmental Policy as Global Assemblage

Understanding power in a Foucauldian sense is integral in the way that he says that we should be attentive to the varied and capillary forms of state action, and this is of particular relevance with contemporary global environmental governance, and specifically in Africa. Rather than seeing the state exercise sovereign power and other actors exert various forms of power, in this book I understood power as pervasive and enacted by the relations of these many actors and not within any specific state confines, boundaries, or known hierarchical manner. Foucault advised focusing on micro-level practices and technologies of governance, and seeking to account for the ways in which binaries such as public and private, state and civil society, national and international are produced, reified, transgressed, eroded, and reformed.[14] Productive power is that which is most prevalent in this book as it is the kind of power we exercise over ourselves based on our knowledge of how we fit into society. We discipline ourselves on the basis of messages we get from society—and CBRNM changes these messages. We try to be normal by disciplining ourselves in the absence of threats or punishment—however those threats remain. With CBNRM we see specifically with several discussions of corruption that the MET is very hesitant to step in, and while it maintains its oversight role to exert productive power from a distance, rarely, or perhaps never does it enact its

[14] Foucault, *Society Must Be Defended*.

sovereign ability to punish through the penal system or de-gazetting conservancies as was suggested would actually be good by several of my interlocutors.

Unlike sovereign power, which is exercised through physical punishments and reward, Foucault tells us productive forms of power are exercised through surveillance and knowledge which we see with the development of the thick CBNRM web composed of community members, conservancy councils, game guards, NGOs, and more. They form a surveillance assemblage which produces knowledge that can be used in a variety of ways—in this sense CBNRM can be said to perform some of the same duties as the panopticon—providing a gaze that has people know they could be watched and thus regulate their own self-discipline. Knowledge is also used to produce power by dissemination of academic and professional/expert knowledge on wildlife conservation and the benefits that can come from incurring the costs that come with living alongside wildlife populations.

While power and its forms are central to existing global assemblage approaches in International Relations, one substantial divergence of my approach is the policy-oriented nature of it. Power remains important, but the many ways it has been theorized by scholars over the years are less important than understanding how this complex policy which I suggest is an iteration of global environmental governance came together, what actions of which actors led to its emergence, and how it has shifted over time. As discussed, policies never exist as one complete thing time immemorial, but instead are always subject to multiple evolving interpretations, enactments, and very often reforms or discontinuation which is where we see power enacted by various actors. For the present study I examine who has the ability to agenda set, and what impacts this agenda has on others involved. Moreover, who has the capacity to keep certain things off the agenda whether this be climate change or various forms of hunting versus photographic tourism. In this book the assemblage approach enables me to analyse how this played out in relation to CBNRM in a number of ways. One of the most obvious of these is with the types of wildlife that are conserved. While predators and elephants are the most dangerous and destructive for local human populations, they tend to be those most highly desired by tourists and global wildlife conservation movements and as a result their conservation takes precedence over other more locally desirable animals which demonstrates the productive power of those in the Global

North/predominantly the West to have their interests realized over local populations.

This was also demonstrated with the types of conservation that took place, specifically as outlined in the discussion of consumptive versus non-consumptive conservation which sees ecotourism win out as well over trophy hunting which can be preferred by many on the ground as more revenue can be garnered with less overhead, resources, and infrastructure needed. I also see the way climate change discussions have largely been left out of the debate serves to benefit ecotourists and those who benefit from them, as that form of conservation is deemed non-consumptive while ignoring the many ways ecotourism can negatively impact the environment and thus keeping discussions of climate change off the agenda is accomplished by certain actors who benefit from omission of this part of the conversation.

These examples all show the ways in which the bargaining power of foreign capital often outweighs the public power of the state or local communities. At the same time the state is enlisting actors in order to provide them with a means by which to exert sovereign power in order to make decisions affecting public interests in various regions which drives policies in certain directions. This foreign capital creates productive power in which NGOs and organizations such as the GCF are able to accomplish effective action within conservancies, often with little oversight. As a result, NGOs are often both an effect of productive power that is enabled by CBNRM and producers of this productive power. Often then we see these actors in some ways produce public power and govern through the ability to channel global capital and in ways enhance the compulsory or sovereign power of the state while at the same time in some ways limiting it.

Global assemblage approaches aim to analyse the politics involved in complex global phenomena, and generally do not lend themselves to policy discussions or prescriptions. However, as my intended audience are both academic and practitioner communities, I draw conclusions to further our empirical understanding of CBNRM as an iteration of global environmental governance, as well as analyse its impact on power and politics. I also further unpack how these politics and the social interactions of actors impact policy frameworks as well as the need to at times wade through the politics to make policy decisions that will best realize the three pillars of CBNRM.

Much of the recent literature on CBNRM points to the many partnerships and the network of actors that have enabled this form of resource governance.[15] Many of these scholars however view the process of devolution and corresponding inroads for a number of actors at the community level largely as negative.[16] CBNRM was intended to empower communities, and partners were supposed to work with conservancies up to the point where they could become fully self-sustaining. While many conservancies have now become financially independent, and in that sense are self-sustaining, the nature of CBRNM means that Namibian communities are intimately imbricated in broader global power relations and cannot be understood outside of them. As such, they also rely on ongoing partnership at a variety of levels to not only ensure continued success, but importantly evolve in response to new challenges. Climate change has proven to be the biggest ongoing challenge and it is through novel partnerships the programmes have been able to respond in order to continue to adapt and evolve. While much of this evolution has been positive, there remains a great deal to be accomplished with the third pillar of community empowerment, and continued evolution ought to see increasing numbers of Namibians, not only hired to work for the NGOs which are central to CBNRM, but also to be in positions of power. Their voices should be central in decision making and resource allocation so they can develop the kinds of skills and experience that will allow them to leverage their existing local cultural and historical experiences, in order to engender the type of ownership that is so often lacking in development initiatives.

Many existing studies have negatively viewed the ongoing power that foreign actors are able to exercise on the ground whether through funding to NGOs,[17] by supporting certain conservation activities in specific conservancies,[18] or by providing resources for certain groups or regions based on desires emanating from outside the country.[19] While it can be assumed that all involved do not have altruistic motives, part of the benefit

[15] Ndeinoma, Wiersum, and Arts, "The Governance of Indigenous Natural Products in Namibia."

[16] Duffy et al., "Why We Must Question the Militarisation of Conservation"; Schnegg and Kiaka, "Subsidized Elephants"; Schnegg, Bollig, and Linke, "Moral Equality and Success of Common-Pool Water Governance in Namibia."

[17] Stoldt et al., "Transfrontier Conservation Areas and Human-Wildlife Conflict."

[18] Ndeinoma, Wiersum, and Arts, "The Governance of Indigenous Natural Products in Namibia."

[19] Duffy, "The Potential and Pitfalls of Global Environmental Governance"; Brockington and Scholfield, "Expenditure by Conservation Nongovernmental Organizations in Sub-Saharan Africa."

of the way CBNRM has developed is that actors seeking to realize their own goals can, and often do, contribute to synergistic outcomes. Thus, the evolving assemblage of actors has contributed to a necessary evolution of the programme over several decades. As we know that actors involved do not all share similar interests, CBNRM must be understood as a product of evolutions in global environmental governance that is composed of an assemblage of actors rather than a domestic policy prescription.

As has been discussed in this book, rather than conservancies increasingly becoming self-sufficient and thus decreasing the need for partnerships and the number of actors involved in CBNRM, climate change has actually contributed to a proliferation of actors and the partnerships these entail. Thus, while communities continue to work to educate and train their populations and expand their ability to realize certain interests, they do so alongside a growing and constantly changing assemblage of actors which includes NGOs, multilateral organizations, private interests, as well as a number of emerging hybrid entities such as the Green Climate Fund. Inevitably, powerful international actors who command global financing streams and who are most often supported by large parts of the international community are able to exercise their will and it is most often their goals which are realized. To be sure there can also be spillover effects where communities also benefit, or that at times goals are not synergistic. However, it is to say that as climate change has increasingly threatened CBNRM, rather than communities gaining autonomy and becoming empowered, what we are often witnessing is more of the same. In this case, business as usual means communities that were intended to gain increased rights and responsibilities are often limited by the will of the international community which can hinge on volatile financial flows and shifting fashions. This is not to say communities do not possess agency, and in many cases even increased agency as a result of CBNRM. However, the agency and autonomy of communities are limited in their need to work with global partners who exert productive power in order to seek to responsibilize said communities to pursue certain conservation and developmental pathways in order to earn rewards as well as to avoid penalties.

Analysing the local/global nature of CBNRM and understanding that it is a much more complex set of processes than a devolution of power from the central government to local communities becomes integral in order to account for these competing sets of interests and desires that can be divergent but can also be synergistic and productive. In general, all actors involved share the goals of fighting climate change, achieving sustainable development, and protecting wildlife and their habitats. Thus, working towards these broad programme goals is what is discussed and espoused by key actors involved. However, increased populations of certain wildlife species can be dangerous and destructive for local communities. While communities might support the conservation of giraffe, zebras, and various types of antelope, these are rarely those species most sought after by tourists as well as not those which global conservation actors seek to protect most. Furthermore, while building a solar power plant in one conservancy can produce a wealth of benefits for one conservancy and can be pointed to as a measurable win for the international community and even the Namibia state, other communities around Namibia will likely experience no benefit at all, and may even see their own revenues decline if tourism shifts to greener lodges. These are just a few examples of the ways in which achieving certain goals can benefit a number of actors but might simultaneously impede the realization of other alternative and for other actors. This complexity makes the assemblage approach useful for analysing the constantly shifting nature of this governance arrangement and its multifarious actors.

While I have outlined many of the benefits that ongoing partnerships can and have entailed, it is clear that these benefits are not always enjoyed by African communities. However, acknowledging ongoing power relations is a pivotal starting point in order to adapt policies to ensure communities remain at the centre of resource governance, and to ensure that they are increasingly central in decision-making processes. Beyond this, it is integral to ensure that no individual or small group from any one community is heard above the rest, but instead that decision making is conducted through a democratic process. While the democratic process generally tailors outcomes to the will of the majority, it is important to develop a robust system of checks and balances in order to avoid the tyranny of the majority as cautioned by Alexis de Toqueville.[20] For conservancies this means ensuring consistent, free, and fair elections for conservancy councils based on the constitutions laid out and agreed to by each conservancy. It also means that any revenue sharing plan must remain transparent and adhered to. In order to ensure this, regular AGMs must be held as outlined in the 1996 Nature Conservation Amendment Act, and if these are ignored by any conservancy council, action should be

[20] Alexis de Tocqueville, "Tyranny of the Majority," in *Democracy in America* (New York: Saunders and Otley, 1835), 255–68.

taken against them. While CBNRM is an example of the state exercising productive power in order to rule from a distance, its coercive functions remain ever-present and are important oversight mechanisms to prevent mismanagement and corruption. In order for communities to be empowered and to remain at the centre of development, politics must be analysed both within communities and between communities and other actors. Failure to do so will result in the same type of ongoing failure that has been seen for decades with regard to development efforts in general, and resource governance in particular, in many southern African countries.

International Community-Based Natural Resource *Governance*?

The aim of this book is to inform both academic and practitioner-oriented communities. For academics, analysing CBNRM as a complex governance assemblage enables us to better comprehend the way power is reorganized in a multitude of complex ways. For a number of years scholars have analysed evolutions in global governance that have seen a disaggregation of the state and a proliferation of sub-state, supra-state, and non-state actors. Death argues that "over the past three decades the lure of the state has also been challenged by transnational networks, sub- and supranational institutions, private actors, and civil society groups."[21] As a result, the state is no longer the sole or pre-eminent actor in international relations and CBNRM provides an effective empirical example of the way in which these processes continue to develop. The state has taken a step back to take a less direct/coercive approach to governing resources; however, it continues to exercise control as the sovereign over its subjects in a number of important ways. The implementation of the Nature Conservation Amendment Act was intended to responsibilize and discipline citizens for the sustainable management of their local resources, as well as their own development and governance. With this has come a confluence of partnerships with actors that have continued to proliferate, resulting in a complex governance arrangement. This book's methodological approach enables us to account for and unpack this multitude of heterogeneous actors and their relation to communities, climate change, and wildlife among a plethora of other units of analysis. It does so by approaching this phenomenon with a

[21] Carl Death, *The Green State in Africa*, Yale Agrarian Studies (New Haven: Yale University Press, 2016), 21.

flattened ontology to allow us to account for these novel approaches to governance which exceed the limits of traditional state-centric theories. This approach provides the tools required to analyse and unpack the multi-scalar connections that emerge between units of analysis as well as to understand how power is produced between them.

While this book has unpacked the complexities of the politics of CBNRM to dig through the multitude of constituent parts that compose this assemblage, the approach also has limitations as the entirety of any assemblage cannot be caught in any single analysis or time. Instead, I have traced the power flows that enable the implementation and ongoing management of Namibia's CBNRM policy. Analysing forms of global environmental governance enables us to account for the ever-evolving nature of policies and practices as well as the ways in which they are operationalized and materialize in unique ways in diverse settings. This occurs as power is devolved from the state, and then employed in various regions by a different set of actors who are operating under unique and ever-changing circumstances. Thus, the policy cannot be said to have X outcome, and Y impact on Namibia, or on the people of the Kunene region. Instead, like the state itself, CBNRM is a set of policies and practices that are continuously changing and being impacted by diverse forces and events, while also simultaneously impacting everything around it.

For practitioner communities the question is often some version of, *is CBNRM effective, and how can it be improved?* There is no simple answer to this, but my research has demonstrated that this form of resource governance has undeniably provided a number of important benefits to Namibians and can continue to do so. These benefits however must be understood alongside the ongoing challenges that come with them as well. Specifically, in relation to the global nature of CBNRM which enables a foothold in the country for outside actors to realize their goals within communities. While the CBNRM programme was intended for communities to become self-sustaining, a multitude of complex challenges—of which climate change is central—have contributed to a constantly renewing situation in which ongoing partnerships are engaged in. Thus, rather than CBNRM serving to empower communities and provide them with ownership over their own development, communities continue to be ever reliant on the will of the international community as well as to fluctuations in developing academic trends and discourse. As a result, CBNRM might better be referred to as *International* Community-Based Natural Resource *Governance*, as this form of resource governance is indivisible from global environmental governance and is also an important layer of governance beyond resource management. This is seen with the increasing aversion to trophy hunting globally, as well as to renewed efforts to fight climate change which has led to an expansion in partnerships to provide solar power generation capacities for some conservancies. It is also seen in the way Namibians have become produced through the discourse of climate change as it now infiltrates their almost every conversation and action. This is certainly not an apolitical phenomenon but one that has infiltrated through their renewed connectivities to global actors. While in and of itself neither of these is inherently problematic, it means that conservancies risk being constantly subject to shifts in funding streams and priorities as a result of changing global fashions. Beyond that, there remains the greater risk that CBNRM in general, and Namibia in particular, might lose their appeal within development discourse as well as a form of, and popular region for, ecotourism. COVID-19 has already demonstrated the severity with which exogenous shocks can impact the viability of the programme. While the pandemic is sure to subside and tourism levels are already increasing once again, it is far from certain the industry will remain viable in long term for the people of Namibia—particularly in the face of growing climate challenges.

All of this is not to say that ongoing partnerships are nefarious, or that they are not creating some good, as certainly they are, and I have outlined many of the positive benefits created from CBNRM throughout this book. However, it is clear that the lack of community empowerment that was pointed to by much of the more critical literature has continued—albeit in different ways. We see ways CBNRM has evolved as many of the interview respondents from the research for this book were quick to assert that they make their own decisions and have ownership over local conservation and development initiatives. Nonetheless, they remain locked in a pattern of seeking technical assistance and resources from global partners who increasingly allocate these based on the types of initiatives they favour on the ground which undermines the autonomy of communities and their decision making while demonstrating the ongoing productive power of global capital.

As benefits derived from CBNRM and managed through conservancy councils decreased throughout the 2010s there was a parallel contraction in support for the programmes[22] as well as increasing accusations of local-level corruption,[23] and suggestions that communities were being used to prop up the tourism industry for foreign travellers.[24] These sentiments mixed with increases in human-wildlife conflict were leading to decreasing political support for the programme which had once held widespread backing.[25] As the drought has stretched on year after year, however, and as many traditional ways of life—largely consisting of various forms of subsistence farming for over 60 per cent of Namibians—have become increasingly difficult for many, and nearly impossible for some, there has been a renewed sense of the necessity of conservation among many.[26] These sentiments were outlined by one police officer from the town of Bersig in the Torra Conservancy who said,

> There have always been political issues with the conservancy like anywhere else, but they are not bad, and we were able to vote for a new council in December [2019] which makes us at least feel like our voices are heard. If we cannot vote for a good council at some point, it is a reflection on us, not a fault with the programme. Conservation now is important especially with the drought, even though the drought makes it more difficult.[27]

His sentiments encompass many of the ongoing challenges facing the programme, but also acknowledge that now more than ever conservation is important, as their land and resources are all they have left in some cases.

These kinds of sentiments were echoed by many throughout Kunene Region in a number of conservancies as people first discussed the hardships faced by the drought, but then re-iterated the ongoing need to continue working for conservation. One interview respondent posited "one of the most important things we get is conservation. Before no one cared about it and the animals were disappearing and their habitat being

[22] Shilongo, Sam, and Simuela, "Using Incentives as Mitigation Measure for Human Wildlife Conflict Management in Namibia."

[23] Hoole, "Place-Power-Prognosis."

[24] Schnegg and Kiaka, "Subsidized Elephants."

[25] Shilongo, Sam, and Simuela, "Using Incentives as Mitigation Measure for Human Wildlife Conflict Management in Namibia."

[26] Mwakikagile, *Namibia*.

[27] Interview, Police officer, Bersig, Torra Conservancy, February 4, 2020.

destroyed. Now people take pride in protecting them and we all do it together as a community."[28] While none of the interlocutors I spoke with suggested that CBNRM was any sort of panacea, even several farmers acknowledged that while wildlife conservation can result in attacks on their livestock, overall the programme was beneficial, and if rain does not come, tourism incomes based on wildlife might be all they have left.[29]

Child and Barnes in referring to Charles Tilly write, "The conditions for democracy are similarly onerous, and the history of mankind shows how rarely these conditions have been met (Tilley 2007):[30] the light of democracy flickering briefly around the times of Greek and Roman civilization, but democracy only emerging more generally in the last two centuries."[31] While Western democracies have been forged over centuries of conflict and social development, African states have not followed the same trajectory.[32] Instead, electoral democracy in the Weberian state is something that was implanted by departing colonial regimes who left a period of oppression, conflict, and challenges in their wake. As a result, democracy is often pointed to as something that has failed to succeed in many African contexts. CBNRM has attempted to establish democratic systems at the local level on extremely short timelines. As such, this form of resource governance has met a great deal of challenges across Southern Africa and has failed to realize many of its intended goals. However, by focusing on the continued development of local-level democratic institutional development and community empowerment, attaining the three pillars of CBNRM remains possible even in the face of growing climate challenges.

[28] Focus Group 1, Anabeb Conservancy, February 7, 2020.

[29] Interview, Farmer 2, #Khoadi !Hoas Conservancy, February 10, 2020.

[30] Charles Tilly, *Democracy* (Cambridge University Press, 2007).

[31] Child and Barnes, "The Conceptual Evolution and Practice of Community-Based Natural Resource Management in Southern Africa," 284.

[32] William Reno, *Warlord Politics and African States* (Lynne Rienner Publishers, 1998).

Annex I: Interviews Cited

1. Lara Diez, Nyae Nyae Development Foundation of Namibia, January 10 2020.
2. Dr Ekkehard Klingelhoeffer, Head of Department—Wildlife Management and Ecotourism, University of Namibia, January 13, 2020.
3. Maxi Pia Louis, Director, Namibia Association of CBNRM Support Organizations, January 14, 2020.
4. Marina Tavolaro, PhD candidate in Biological Sciences at University of Cape Town, January 15, 2020
5. Uatirohange Tjiuoro, Ministry of Environment and Tourism: Directorate of National Parks and Wildlife, January 20, 2020.
6. Michael Knight, Transboundary Leader for KAZA, WWF, January 23, 2020.
7. John Hazam, Senior Technical Advisor: Namibia Nature Foundation, January 23, 2020.
8. Joni Seager, Professor of Global Studies, Bentley University, January 24, 2020.
9. Kevin Coldrey, Environmental and climate change economist for Environmental-Economics Policy Research Unit & Anchor Environmental, January 24, 2020
10. Dr Chris Brown, CEO Namibia Chamber of Environment, January 24, 2020.

A. Heffernan, *The Global Politics of Local Conservation*, Environmental Politics and Theory,
https://doi.org/10.1007/978-3-031-24177-2

11. Richard Diggle, Business/CBNRM Specialist, WWF, January 24, 2020.
12. Willie Boonzaaier, Programme Director, Integrated Rural Development and Nature Conservation, January 24, 2020.
13. Eric Xaneb, Chairman, Tsiseb Conservancy, January 27, 2020.
14. Hiking guide, Branberg 'White Lady' paintings, January 28, 2020.
15. Sarah Roos, Manager, Brandberg White Lady Lodge, January 28, 2020.
16. David Eiseb, manager, Sorris Sorris Conservancy, January 29, 2020.
17. Benedict Libanda, CEO, Environmental Investment fund of Namibia, January 29, 2020.
18. Traditional Authority, Sorris Sorris Conservancy, January 29, 2020.
19. Chairperson, Sorris Sorris Conservancy, January 29, 2020.
20. Pohamba Shifeta, Minister of Environment and Tourism, January 29, 2020.
21. Adolph Thaniseb, Uisbased/Twyfelfontein, January 30, 2020.
22. Joglinde Touros, Manager, Uisbasen/Twyfelfontein Conservancy, January 30, 2020.
23. Hiking guide, Twyfelfontein engravings (UNESCO World Heritage Site), January 30, 2020.
24. Guide, Damara Living Museum, twyfelfontein, January 30, 2020.
25. Village respondent 1, Uisbasen/Twyfelfontein, January 30, 2020.
26. Village respondent 2, Uisbasen/Twyfelfontein, January 30, 2020.
27. Village respondent 3, Uisbasen/Twyfelfontein, January 30, 2020.
28. Barnabas, tour guide, Petrified Forest, Uisbasen/Twyfelfontein, January 31, 2020.
29. Mine worker, Bersig, February 1, 2020.
30. Village respondent 4, Torra Conservancy, February 1, 2020.
31. Village respondent 5, Torra Conservancy, February 1, 2020.
32. Emil Roman, Manager, Torra Conservancy, February 2, 2020.
33. Village respondent 6, Torra Conservancy, February 2, 2020.
34. Village respondent 7, Torra Conservancy, February 2, 2020.
35. Farmer 1, Torra Conservancy, February 2, 2020.
36. Game officer, Torra Conservancy, February 2, 2020.
37. Village respondent 8, Torra Conservancy, February 2, 2020.
38. Village respondent 9, Torra Conservancy, February 2, 2020.
39. Manager, Damaraland Camp, February 2, 2020.
40. Anna Carizzoni, Tour guide, Matiti tours, February 3, 2020.
41. Manager, Wereldsend research outpost, February 3, 2020.

42. Employee, Wereldsend research outpost, February 3, 2020.
43. Team lead, NGO Team from France, February 3, 2020.
44. Rhino game guard, Torra Conservancy, February 4, 2020.
45. Police officer, Bersig, Torra Conservancy, February 4, 2020.
46. Manager, Palmwag Lodge, February 5, 2020.
47. Uisa, Manager, Sesfontein Conservancy, February 6, 2020.
48. Traditional authority, Sesfontein Conservancy, February 6, 2020.
49. Game guard, Sesfontein Conservancy, February 6, 2020.
50. Tour guide, Sesfontein Conservancy, February 6, 2020.
51. Head Lady, Sesfontein Conservancy, February 6, 2020.
52. Official, Ministry of Rural Economic Development, February 6, 2020.
53. Christine Awaras, Manager, Anabeb Conservancy, February 7, 2020.
54. Godlieb, Office Manager, Anabeb Conservancy, February 7, 2020.
55. Traditional Authority 1, Anabeb Conservancy, February 7, 2020.
56. Traditional authority 2: Anabeb Conservancy, February 7, 2020.
57. Focus group 1, Anabeb Conservancy, February 7, 2020.
58. Focus group 2, Anabeb Conservancy, February 7, 2020.
59. Focus group 3, Anabeb Conservancy, February 7, 2020.
60. Focus group 4, Anabeb Conservancy, February 7, 2020.
61. Manager, Okahirongo Elephant Lodge, February 8, 2020.
62. Secretary, #Khoadi !Hoas Conservancy, February 10, 2020.
63. Andrew Naopina, treasurer, #Khoadi !Hoas Conservancy, February 10, 2020.
64. Albert Kroses, Head of Environmental Services, #Khoadi !Hoas Conservancy, February 10, 2020.
65. Landina, Office Assistant, #Khoadi !Hoas Conservancy, February 10, 2020.
66. Manager, Grootberg Lodge, #Khoadi !Hoas Conservancy, February 10, 2020.
67. Manager, Hoada Campsite, #Khoadi !Hoas Conservancy, February 10, 2020.
68. Farmer 2, #Khoadi !Hoas Conservancy, February 10, 2020.
69. Farmer 3, #Khoadi !Hoas Conservancy, February 10, 2020.
70. Employee: Grootberg Lodge, #Khoadi !Hoas Conservancy, February 10, 2020.
71. Employee: Hoada campsite, #Khoadi !Hoas Conservancy, February 10, 2020.

72. Etosha Park Warden, Ministry of Environment and Tourism satellite office, February 12, 2020.
73. Lauren Pfeiffer, Assistant to Director of the Cheetah Conservation Fund, February 13, 2020.
74. Dr Margaret Jacobson, co-founder IRDNC, co-director, trustee, mentor, Conservancy Safaris Namibia, Swakopmnud, February 17, 2020.
75. Andrew Malherbe, COO, Save the Rhino Trust, Swakopmund, February 18, 2020.
76. Official, Ministry of Environment and Tourism, Swakopmund, February 19, 2020.
77. Gitta Paetzold, CEO, Hospitality Association of Namibia, Swakopmund, February 19, 2020.
78. Brian Jones, Environment and Development Consultant, Windhoek, February 20, 2020.
79. John Pallet, Director, Namibia Environment and Wildlife Society, Windhoek, February 21, 2020.
80. Ruben Portas, Cheetah Research Project. Windhoek, January 14, 2020.

Bibliography

Abrahamsen, Rita. "Assemblages." Routledge Handbook of International Political Sociology, December 1, 2016a.

———. "Security Privatisation, the State, and Development in the Global South." In *The Palgrave Handbook of International Development*, edited by Jean Grugel and Daniel Hammett, 243–57. London: Palgrave Macmillan UK, 2016b.

———. "The Power of Partnerships in Global Governance." *Third World Quarterly* 25, no. 8 September 1, (2004): 1453–67.

Abrahamsen, Rita, and Michael C. Williams. *Security beyond the State: Private Security in International Politics*. Cambridge, UK; New York: Cambridge University Press, 2011.

Abrahamsen, Rita. "Security Privatization and Global Security Assemblages." *The Brown Journal of World Affairs* 18, no. 1 (2011): 171–80.

Acuto, Michele, and Simon Curtis. "Assemblage Thinking and International Relations." In *Reassembling International Theory: Assemblage Thinking and International Relations*, edited by Michele Acuto and Simon Curtis, 1–15. London: Palgrave Macmillan UK, 2014.

Agrawal, Arun, and Clark C. Gibson. "Enchantment and Disenchantment: The Role of Community in Natural Resource Conservation." *World Development* 27, no. 4 (1999): 629–49.

Ahlborg, Helene, Frida Boräng, Sverker C. Jagers, and Patrik Söderholm. "Provision of Electricity to African Households: The Importance of Democracy and Institutional Quality." *Energy Policy* 87 (December 1, 2015): 125–35.

A. Heffernan, *The Global Politics of Local Conservation*, Environmental Politics and Theory,
https://doi.org/10.1007/978-3-031-24177-2

Ashley, Caroline. "The Impacts of Tourism on Rural Livelihoods: Namibia's Experience." Overseas Development Institute London, UK, February 2000.

Ashley, Caroline, Jon Barnes, Chris Brown, and Brian Jones. "Using Resource Economics for Natural Resource Management: Namibia's Experience," 1997.

Barnett, Michael, and Raymond Duvall. *Power in Global Governance*. Cambridge, United Kingdom: Cambridge University Press, 2004.

Bartelson, Jens. "Three Concepts of Globalization." *International Sociology* 15, no. 2 (2000): 180–96.

Bennett, Jane. "The Agency of Assemblages and the North American Blackout." *Public Culture* 17, no. 3 (September 1, 2005): 445–66.

Berkes, Fikret. "Commons in a Multi-Level World." *International Journal of the Commons* 2, no. 1 (November 1, 2007): 1–6.

Berkes, Fikret, Gary P. Kofinas, and F. Stuart Chapin. "Conservation, Community, and Livelihoods: Sustaining, Renewing, and Adapting Cultural Connections to the Land." In *Principles of Ecosystem Stewardship: Resilience-Based Natural Resource Management in a Changing World*, edited by Carl Folke, Gary P. Kofinas, and F. Stuart Chapin, 129–47. New York, NY: Springer New York, 2009.

Best, Jacqueline. "Technocratic Exceptionalism: Monetary Policy and the Fear of Democracy." *International Political Sociology* 12, no. 4 (December 1, 2018): 328–45.

———. "Why the Economy Is Often the Exception to Politics as Usual." *Theory* 24, no. 4 (2007): 87–109.

Bollig, Michael. "Towards an Arid Eden? Boundary Making, Governance and Benefit Sharing and the Political Ecology of the 'New Commons' of Kunene Region, Northern Namibia." *International Journal of the Commons* 10, no. 2 (2016): 771–99.

Bollig, Michael, and Diego Menestrey Schwieger. "Fragmentation, Cooperation and Power: Institutional Dynamics in Natural Resource Governance in North-Western Namibia." *Human Ecology* 42, no. 2 (2014): 167–81.

Bos, Kristine, Duncan Chaplin, and Arif Mamun. "Benefits and Challenges of Expanding Grid Electricity in Africa: A Review of Rigorous Evidence on Household Impacts in Developing Countries." *Energy for Sustainable Development* 44 (June 1, 2018): 64–77.

Boudreaux, Karol. "A New Call of the Wild: Community-Based Natural Resource Management in Namibia Essay." *Georgetown International Environmental Law Review* 20 (2008): 297–336.

Bourdieu, Pierre. "The Forms of Capital." In *Readings in Economic Sociology*, 280–91. John Wiley & Sons, Ltd, 2008.

Bracking, Sarah. "The Anti-Politics of Climate Finance: The Creation and Performativity of the Green Climate Fund: The Anti-Politics of Climate Finance." *Antipode* 47, no. 2 (2015): 281–302.

Bratton, Michael, and Nicholas van de Walle. "Chapter 2: Neopatrimonial Rule in Africa." In *Democratic Experiments in Africa Regime Transitions in Comparative Perspective*. Cambridge University Press, 1997.

Brockington, Dan, and Katherine Scholfield. "Expenditure by Conservation Nongovernmental Organizations in Sub-Saharan Africa." *Conservation Letters* 3, no. 2 (2010): 106–13.

Brockington, Daniel, and James Igoe. "Eviction for Conservation: A Global Overview." *Conservation and Society* 4, no. 3 (July 1, 2006): 424.

Brodie, Bernard, and Frederick Sherwood Dunn. *The Absolute Weapon: Atomic Power and World Order*. Harcourt, Brace and Company, 1946.

Brown, Chris. "Opinion: Ecologist Responds to Guardian Newspaper Article against Trophy Hunting." *Africa Geographic* (blog), January 2, 2019.

Brown, Jessica, and Neil Bird. "Sustainable Natural Resource Management in Namibia." *Development Progress*, 2011.

Buscher, Bram. "Anti-Politics as Political Strategy: Neoliberalism and Transfrontier Conservation in Southern Africa." *Development and Change* 41, no. 1 (2010): 29–51.

Carr, E. H. *The Twenty Years' Crisis, 1919-1939: Reissued with a New Preface from Michael Cox*. Springer, 2016.

Cash, D., W. Adgef, F. Berkes, P. Garden, L. Lebel, P. Olsson, L. Pritchard, and O. Young. "Scale and Cross-Scale Dynamics: Governance and Information in a Multilevel World." *Ecology and Society* 11, no. 2 (2006).

Ceballos, Gerardo, Paul R. Ehrlich, and Peter H. Raven. "Vertebrates on the Brink as Indicators of Biological Annihilation and the Sixth Mass Extinction." *Proceedings of the National Academy of Sciences* 117, no. 24 (June 16, 2020): 13596–602.

Cerny, Philip G. "Paradoxes of the Competition State: The Dynamics of Political Globalization." *Government and Opposition* 32, no. 2 (1997): 251–74.

Chamberlain, Muriel Evelyn. *The Scramble for Africa*. 3rd ed. Seminar Studies. Oxfordshire, England: Routledge, 2009.

Child, Brian. "Origins and Efficacy of Modern Community Based Natural Resources Management (CBNRM) Practices in the Southern African Region." *Snively*, 2003.

———. "The Practice and Principles of Community-Based Wildlife Management in Zimbabwe: The CAMPFIRE Programme." *Biodiversity & Conservation* 5, no. 3 (1996): 369–98.

Child, Brian, and Grenville Barnes. "The Conceptual Evolution and Practice of Community-Based Natural Resource Management in Southern Africa: Past, Present and Future." *Environmental Conservation* 37, no. 3 (2010): 283–95.

CIA World Factbook. "Africa: Namibia — The World Factbook - Central Intelligence Agency," 2020.

City of Cape Town. "Our Shared Water Future. Cape Town's Water Strategy," 2019.

Clifford, Nicholas, Meghan Cope, Thomas Gillespie, and Shaun French. *Key Methods in Geography*. SAGE, 2016.

Coldrey, Kevin, and Jane Turpie. "Climate Change Risk Assessment for Namibia's Communal Conservancies," 2020a.

———. "Climate Change Risk Assessment for Namibia's Communal Conservancies," 2020b.

Collier, Stephen J. "Assemblages and the Conduct of Inquiry." In *Reassembling International Theory: Assemblage Thinking and International Relations*, edited by Michele Acuto and Simon Curtis, 32–38. London: Palgrave Macmillan UK, 2014.

Collomb, Jean-Gael E., Patricia Mupeta, Grenville Barnes, and Brian Child. "Integrating Governance and Socioeconomic Indicators to Assess the Performance of Community-Based Natural Resources Management in Caprivi (Namibia)." *Environmental Conservation* 37, no. 3 (2010): 303–9.

Commission on Global Governance. *Our Global Neighbourhood: The Report of the Commission on Global Governance*. Oxford; New York: Oxford Paperbacks, 1995.

Cooke, Bill, and Uma Kothari. *Participation: The New Tyranny?* Zed Books, 2001.

Cooper, Frederick. *Colonialism in Question: Theory, Knowledge, History*. Berkeley: University of California Press, 2005.

———. "What Is the Concept of Globalization Good for? An African Historians Perspective." *African Affairs* 100, no. 399 (2001): 189–213.

Cornwall, Andrea, and Karen Brock. "What Do Buzzwords Do for Development Policy? A Critical Look at 'Participation', 'Empowerment' and 'Poverty Reduction.'" *Third World Quarterly* 26, no. 7 (October 1, 2005): 1043–60.

Corson, Catherine. "Shifting Environmental Governance in a Neoliberal World: US AID for Conservation." *Antipode* 42, no. 3 (2010): 576–602.

Cruise, Adam, and Sasada Izzy. "Investigation into the Efficacy of Namibia's Wildlife Conservation Model as It Relates to African Elephants (Loxodonta Africana)," 2021.

Dahl, Robert A. "The Concept of Power." *Behavioral Science* 2, no. 3 (1957): 201–15.

Dean, Mitchell. *Governing Societies: Political Perspectives on Domestic and International Rule*. Issues in Society. Maidenhead, England; Open University Press, 2007.

Death, Carl. *The Green State in Africa*. Yale Agrarian Studies. New Haven: Yale University Press, 2016.

Diop, Makhtar. "Foresight Africa 2016: Banking on Agriculture for Africa's Future." *Brookings* (blog), 2016.

Doolittle, Amity A. "Fortress Conservation." In *Encyclopedia of Environment and Society*, 705–705. Thousand Oaks: SAGE Publications, Inc., 2007.

Duffy, Rosaleen. "Global Environmental Governance and the Challenge of Shadow States: The Impact of Illicit Sapphire Mining in Madagascar." *Development and Change* 36, no. 5 (2005): 825–43.

———. "The Potential and Pitfalls of Global Environmental Governance: The Politics of Transfrontier Conservation Areas in Southern Africa." *Political Geography* 25, no. 1 (2006): 89–112.

Duffy, Rosaleen, Francis Massé, Emile Smidt, Esther Marijnen, Bram Büscher, Judith Verweijen, Maano Ramutsindela, Trishant Simlai, Laure Joanny, and Elizabeth Lunstrum. "Why We Must Question the Militarisation of Conservation." *Biological Conservation* 232, no. Complete (2019): 66–73.

Dunn, Elizabeth C. "Standards and Person-Making in East Central Europe." In *Global Assemblages*, 173–93. John Wiley & Sons, Ltd, 2008.

Dyer, Jen, Lindsay C. Stringer, Andrew John Dougill, Julia Leventon, Muleba Nshimbi, Francis Chama, Ackson Kafwifwi, et al. "Assessing Participatory Practices in Community-Based Natural Resource Management: Experiences in Community Engagement from Southern Africa." *Journal of Environmental Management* 137, no. Complete (2014): 137–45.

EIFN. "Our History: EIF Through the Years," n.d.

———. "Partners." Accessed November 23, 2021.

Elden, Stuart. "Missing the Point: Globalization, Deterritorialization and the Space of the World." *Transactions of the Institute of British Geographers* 30, no. 1 (2005): 8–19.

Etosha National Park. "Etosha History." Accessed February 29, 2020.

Fabricius, C., and S. Collins. "Community-Based Natural Resource Management: Governing the Commons." *Water Policy* 9, no. S2 (November 2007): 83–97.

Fairhead, James, Melissa Leach, and Ian Scoones. *Green Grabbing: A New Appropriation of Nature*. Critical Agrarian Studies. London: Routledge, 2013.

FAO. "Namibia at a Glance | FAO in Namibia | Food and Agriculture Organization of the United Nations." Accessed May 7, 2020.

Flightradar24. "Live Flight Tracker - Real-Time Flight Tracker Map." Flightradar24, 2022.

Fortmann, Louise, Emery Roe, and Michel van Eeten. "At the Threshold between Governance and Management: Community-based Natural Resource Management in Southern Africa." *Public Administration and Development* 21, no. 2 (2001): 171–85.

Foucault, Michel. *Discipline and punish: the birth of the prison*. Social theory, 2nd edition. New York: Vintage Books, 1979.

———. "Power/Knowledge: Selected Interviews and Other Writings." Penguin Random House Canada, 1977.

———. *Power/Knowledge: Selected Interviews and Other Writings, 1972-1977*. 1st ed. New York: Vintage, 1980.

———. *Security, Territory, Population: Lectures at the Collège de France, 1977-78.* Houndmills, Basingstoke, Hampshire; Palgrave Macmillan, 2007.

———. *"Society Must Be Defended": Lectures at the Collège de France, 1975-76.* 1st Picador pbk. ed. New York: Picador, 2003.

———. "The Subject and Power." *Critical Inquiry* 8, no. 4 (1982): 777–95.

Fund, Green Climate. "Initial Resource Mobilisation." Text. Green Climate Fund. Green Climate Fund, October 26, 2021.

NACSO. "Funding Partners," October 11, 2015a.

Galvin, Kathleen A., Danielle Backman, Matthew W. Luizza, and Tyler A. Beeton. "African Community-Based Conservancies: Innovative Governance for Whom?" In *Nomad-State Relationships in International Relations*, edited by Jamie Levin, 147–72. Cham: Springer International Publishing, 2020.

Galvin, Kathleen, Tyler Beeton, and Matthew Luizza. "African Community-Based Conservation: A Systematic Review of Social and Ecological Outcomes." *Ecology and Society* 23, no. 3 (September 11, 2018).

Gaodirelwe, Ikanyeng, Moseki Ronald Motsholapheko, and Gaseitsiwe Smollie Masunga. "Community Perceptions of Wildlife Management Strategies and Subsistence Poaching in the Okavango Delta, Botswana." *Human Dimensions of Wildlife* 25, no. 3 (May 3, 2020): 232–49.

Green Climate Fund. "Governing Instrument for the Green Climate Fund," 2012.

———. "Project Lifecycle." Text. Green Climate Fund. Green Climate Fund, March 18, 2020.

Grovogui, S. *Beyond Eurocentrism and Anarchy: Memories of International Order and Institutions.* Springer, 2016.

Grovogui, Siba N. "Regimes of Sovereignty: International Morality and the African Condition." *European Journal of International Relations* 8, no. 3 (2002): 315–38.

Hall. *Emergence of Private Authority in Global Governance.* Cambridge Studies in International Relations; 85. Cambridge, Cambridge; New York: Cambridge University Press, 2002.

Hall, Rodney Bruce, and Thomas J. Biersteker, eds. *The Emergence of Private Authority in Global Governance.* Cambridge: Cambridge University Press, 2003.

Hardin, Garrett. "The Tragedy of the Commons." *Science* 162, no. 3859 (1968): 1243–48.

Heffernan, Andrew. "Accounting for Climate Change in CBNRM: Reflections on Wildlife Conservation in Namibia." *Journal of Southern African Studies*, 2022a.

———. "Development, Conservation, Empowerment: The Trilemma of Community-Based Natural Resource Management in Namibia." *Environmental Management*, 2022b.

———. "El discurso oficial de las autoridades marroquíes y sus aliados en la ecologización de sus políticas públicas: modernización del autoritarismo, inserción internacional y lucha contra el cambio climático." *Relaciones Internacionales* 0, no. 42 (September 30, 2019): 157–73.

———. "Paralympic Assemblages Globalizing International Relations: An Autoethnographic Account of Global Politics at the Paralympic Games." *Sport in Society* 0, no. 0 (June 5, 2020a): 1–14.

———. "The Globality of Community-Based Conservation in Namibia – Centre for International Policy Studies." 2020b.

Hoole, Arthur Frederick. "Place-Power-Prognosis: Community-Based Conservation, Partnerships, and Ecotourism Enterprises in Namibia." *International Journal of the Commons* 4, no. 1 (2009): 78–99.

Huggins, Chris. "Land-Use Planning, Digital Technologies, and Environmental Conservation in Tanzania." *The Journal of Environment & Development* 27, no. 2 (June 1, 2018): 210–35.

Hulme, David, and Marshall Murphree. *African Wildlife and Livelihoods: The Promise and Performance of Community Conservation*. Great Britain: Woolnough, Irthlingborough, 2001.

Hulme, David. "Communities, Wildlife and the 'New Conservation' in Africa." *Journal of International Development* 11, no. 2 (1999): 277–85.

Humphreys, Macartan, Jeffrey Sachs, and Joseph E. Stiglitz. *Escaping the Resource Curse*. Columbia University Press, 2007.

Igoe, Jim. "Consume, Connect, Conserve: Consumer Spectacle and the Technical Mediation of Neoliberal Conservation's Aesthetic of Redemption and Repair." *Human Geography* 6, no. 1 (March 1, 2013): 16–28.

"Integrated Rural Development and Nature Conservation - Our Team." Accessed November 22, 2021.

"Integrated Rural Development and Nature Conservation - Partners." Accessed November 22, 2021.

IPCC. "Summary for Policymakers — Global Warming of 1.5 °C," 2018.

IRDNC. "Tribute to Garth Owen-Smith, African Conservationist (Commentary)." Mongabay Environmental News, 2020.

Jacobsohn, Margaret. *Life Is Like a Kudu Horn: A Memoir*. Jacana Media, 2019.

Jacobson, P. J., and K. M. Jacobson. "Hydrologic Controls of Physical and Ecological Processes in Namib Desert Ephemeral Rivers: Implications for Conservation and Management." *Journal of Arid Environments*, Deserts of the World: Namib Desert: 50 years of research in a hyperarid desert, 93 (June 1, 2013): 80–93.

Jacquet, Jennifer, and Nicolas Delon. "The Values behind Calculating the Value of Trophy Hunting: The Values Behind Trophy Hunting." *Conservation Biology* 30, no. 4 (August 2016): 910–11.

Jones, Brian. "The Evolution of a Community-Based Approach to Wildlife Management at Kunene, Namibia." *African Wildlife and Livelihoods: The Promise and Performance of Community Conservation*, 2001, 160–76.

———. "The Evolution of Namibia's Communal Conservancies." In *Community Rights, Conservation and Contested Land*, 119–33. Routledge, 2010.

Jones, Brian, and Marshall Murphree. "The Evolution of Policy on Community Conservation in Namibia and Zimbabwe." *African Wildlife and Livelihoods: The Promise and Performance of Community Conservation*, 2001, 38–58.

Jones, Brian T. B. "Policy Lessons from the Evolution of a Community-Based Approach to Wildlife Management, Kunene Region, Namibia." *Journal of International Development* 11, no. 2 (1999a): 295–304.

Jones, Brian TB. *Community-Based Natural Resource Management in Botswana and Namibia: An Inventory and Preliminary Analysis of Progress*. IIED, 1999b.

———. "Policy Lessons from the Evolution of a Community-Based Approach to Wildlife Management, Kunene Region, Namibia." *Journal of International Development: The Journal of the Development Studies Association* 11, no. 2 (1999c): 295–304.

Jones, Brian, and L. Chris Weaver. "CBNRM in Namibia: Growth, Trends, Lessons and Constraints." In *Evolution and Innovation in Wildlife Conservation*. Routledge, 2008.

Kalvelage, Linus, Michael Bollig, Elke Grawert, Carolin Hulke, Maximilian Meyer, Kennedy Mkutu, Marie Müller-Koné, and Javier Diez. "Territorialising Conservation: Community-Based Approaches in Kenya and Namibia." *Conservation and Society* 19, no. 4 (2021): 282–93.

Kang, Susan. "What the Documents Can't Tell You: Participant Observation in International Relations." *PS: Political Science & Politics* 50, no. 1 (January 2017): 121–25.

Katz-Rosene, Ryan, and Matthew Paterson. *Thinking Ecologically about the Global Political Economy*. London: Routledge, 2018.

Kaya, Hassan O, and Yonah N Seleti. "African Indigenous Knowledge Systems and Relevance of Higher Education in South Africa." *The International Education Journal: Comparative Perspectives* 12, no. 1 (2013): 30–44.

Keck, Margaret E. *Activists beyond Borders: Advocacy Networks in International Politics*. Ithaca, N.Y.: Cornell University Press, 1998.

Keohane, Robert O., and Marc A. Levy. *Institutions for Environmental Aid: Pitfalls and Promise*. Global Environmental Accords. Strategies for Sustainability. Cambridge, Mass: MIT Press, 1996.

Kepe, Thembela, and Ruth Hall. "Land Redistribution in South Africa: Towards Decolonisation or Recolonisation?" *Politikon* 45, no. 1 (January 2, 2018): 128–37.

Kepe, Thembela, Rachel Wynberg, and William Ellis. "Land Reform and Biodiversity Conservation in South Africa: Complementary or in Conflict?" *International Journal of Biodiversity Science & Management* 1, no. 1 (March 1, 2005): 3–16.

Kirchherr, Julian, and Katrina Charles. "Enhancing the Sample Diversity of Snowball Samples: Recommendations from a Research Project on Anti-Dam Movements in Southeast Asia." *PLOS ONE* 13, no. 8 (August 22, 2018): e0201710.

Kitzinger, Jenny. "The Methodology of Focus Groups: The Importance of Interaction between Research Participants." *Sociology of Health & Illness* 16, no. 1 (1994): 103–21.

Koot, Stasja. "The Limits of Economic Benefits: Adding Social Affordances to the Analysis of Trophy Hunting of the Khwe and Ju/'hoansi in Namibian Community-Based Natural Resource Management." *Society & Natural Resources* 32, no. 4 (April 3, 2019): 417–33.

Koot, Stasja, Paul Hebinck, and Sian Sullivan. "Science for Success—A Conflict of Interest? Researcher Position and Reflexivity in Socio-Ecological Research for CBNRM in Namibia." *Society & Natural Resources* 0, no. 0 (May 21, 2020): 1–18.

Krawczyk, Kelly. "The Relationship Between Liberian CSOs and International Donor Funding: Boon or Bane?" *VOLUNTAS: International Journal of Voluntary and Nonprofit Organizations* 29, no. 2 (April 1, 2018): 296–309.

LaFranchi, C., and C. Ashley. "Livelihood Strategies of Rural Households in Caprivi: Implications for Conservancies and Natural Resource Management." Working Paper. Directorate of Environmental Affairs Ministry of Environment and Tourism, Windhoek, Namibia, 1997.

Larner, Wendy. "Neo-Liberalism: Policy, Ideology, Governmentality." *Studies in Political Economy*, no. 63 (2000): 5–25.

Larner, Wendy, and William Walters. "Globalization as Governmentality." *Alternatives: Global, Local, Political* 29, no. 5 (2004): 495–514.

Leach, Melissa, Robin Mearns, and Ian Scoones. "Environmental Entitlements: Dynamics and Institutions in Community-Based Natural Resource Management." *World Development* 27, no. 2 (1999): 225–47.

Leal Filho, Walter. *Experiences of Climate Change Adaptation in Africa*. Climate Change Management. Springer Berlin Heidelberg, 2011.

Lendelvo, S. M., M. Pinto, and S. Sullivan. "A Perfect Storm? The Impact of COVID-19 on Community-Based Conservation in Namibia." *Namibian Journal of the Environment* 4 B (July 1, 2020): 1–15.

Lenggenhager, Luregn. *Ruling Nature, Controlling People: Nature Conservation, Development and War in North-Eastern Namibia since the 1920s*. African Books Collective, 2018.

Levy, Jack S. "Case Studies: Types, Designs, and Logics of Inference." *Conflict Management and Peace Science* 25, no. 1 (February 2008): 1–18.

Li, Tania Murray. "Practices of Assemblage and Community Forest Management." *Economy and Society* 36, no. 2 (May 1, 2007): 263–93.

Lubilo, Rodgers, and Paul Hebinck. "'Local Hunting' and Community-Based Natural Resource Management in Namibia: Contestations and Livelihoods." *Geoforum* 101 (2019): 62–75.

Mamdani, Mahmood. *Citizen and Subject: Contemporary Africa and the Legacy of Late Colonialism. Citizen and Subject.* Princeton University Press, 2018.

Manji, Ambreena S. *The Politics of Land Reform in Africa: From Communal Tenure to Free Markets.* Zed Books, 2006.

Mann, Geoff, and Joel Wainwright. *Climate Leviathan: A Political Theory of Our Planetary Future.* Verso Books, 2020.

Mannetti, Lelani M., Thomas Göttert, Ulrich Zeller, and Karen J. Esler. "Expanding the Protected Area Network in Namibia: An Institutional Analysis." *Ecosystem Services* 28, no. Part B (2017): 207–18.

Mannetti, Lelani M. "Identifying and Categorizing Stakeholders for Protected Area Expansion around a National Park in Namibia." *Ecology and Society* 24, no. 2 (July 1, 2019): 5.

Mbembe, Achille. "Provisional Notes on the Postcolony." *Africa* 62, no. 1 (January 1, 1992): 3–37.

McCann, Eugene, and Kevin Ward. *Mobile Urbanism: Cities and Policymaking in the Global Age.* U of Minnesota Press, 2011.

Meadows, Donella, Jorgen Randers, and Dennis Meadows. *Limits to Growth: The 30-Year Update.* Chelsea Green Publishing, 2004.

Melber, Henning. "Colonialism, Land, Ethnicity, and Class: Namibia after the Second National Land Conference." *Africa Spectrum* 54, no. 1 (April 1, 2019): 73–86.

Menestrey Schwieger, Diego A. "An Ethnographic Analysis of the Role of Power in Institutional Arrangements: Borehole Cost Recovery within a Pastoral Community in North-Western Namibia." *Environmental Policy and Governance* 25, no. 4 (2015): 258–69.

Mercer, Claire. "Performing Partnership: Civil Society and the Illusions of Good Governance in Tanzania." *Political Geography* 22, no. 7 (2003): 741–63.

MET. "Nature Conservation Amendment Act," 1996.

Methmann, Chris Paul. "'Climate Protection' as Empty Signifier: A Discourse Theoretical Perspective on Climate Mainstreaming in World Politics." *Millennium* 39, no. 2 (2010): 345–72.

MET/NACSO. "The State of Community Conservation in Namibia: Annual Report 2019: A Review of Communal Conservancies, Community Forests and Other CBNRM Initiatives." Ministry of Environment and Tourism, 2021.

Moore, Lorraine. "The Neoliberal Elephant: Exploring the Impacts of the Trade Ban in Ivory on the Commodification and Neoliberalisation of Elephants." *Geoforum* 42, no. 1 (2011): 51–60.

Morini, Simone (OCCP). "FAO Namibia Newsletter, 1st Semester 2020 - Issue #1," 2020, 10.

Mwakikagile, Godfrey. *Namibia: Conquest to Independence: Formation of a Nation.* 1st edition. Dar es Salaam: New Africa Press, 2015.

NACSO. "Members." NACSO, 2015b.

———. "Resources & Publications: State of Community Conservation." NACSO. Accessed May 7, 2020.

Naidoo, Robin, L. Chris Weaver, Richard W. Diggle, Greenwell Matongo, Greg Stuart-Hill, and Chris Thouless. "Complementary benefits of tourism and hunting to communal conservancies in Namibia." *Conservation Biology* 30, no. 3 (2016): 628–38.

Freedom House. "Namibia: Freedom in the World 2020 Country Report." Accessed September 22, 2022.

Namibia Tourism Board. "Namibia Tourism Satellite Account," 2021.

NASA Earth Observatory. "Drought Threatens Millions in Southern Africa." Text. Article. NASA Earth Observatory, December 16, 2019.

Nattrass, Nicoli. "Conservation and the Commodification of Wildlife in the Anthropocene: A Southern African History." *South African Historical Journal* 0, no. 0 (April 20, 2021): 1–22.

"Nature_Conservation_Amendment_Act.Pdf." Accessed March 18, 2020.

Ndeinoma, Albertina, K. Freerk Wiersum, and Bas Arts. "The Governance of Indigenous Natural Products in Namibia: A Policy Network Analysis." *Environmental Management* 62, no. 1 (July 1, 2018): 29–44.

Nhongo, Kaula. "Namibia Hikes Key Rate by Most in 15 Years to Temper Inflation - BNN Bloomberg." BNN, June 15, 2022.

Niang, Amy. *The Postcolonial African State in Transition: Stateness and Modes of Sovereignty.* Rowman & Littlefield, 2018.

Nuulimba, Karine, and Julie J Taylor. "25 Years of CBNRM in Namibia: A Retrospective on Accomplishments, Contestation and Contemporary Challenges." *Journal of Namibian Studies: History Politics Culture* 18 (December 16, 2015): 89–110.

Ong, Aihwa. *Neoliberalism as Exception Mutations in Citizenship and Sovereignty.* Durham [N.C.]: Duke University Press, 2006.

Ostrom, Elinor. "Coping with the Tragedy of the Commons," 1999.

———. *Governing the Commons: The Evolution of Institutions for Collective Action.* The Political Economy of Institutions and Decisions. r: Cambridge University Press, 1990.

Otto, Jonathan, Charles Zerner, John Robinson, Richard Donovan, Marianne Lavelle, R. Villarreal, Nick Salafsky, et al. *Natural Connections: Perspectives In Community-Based Conservation.* Edited by David Western and Michael Wright. 2 edition. Washington, D.C: Island Press, 1994.

Owa, Masumi. "Revisiting the Paris Declaration Agenda – an Inclusive, Realistic Orientation for Aid Effectiveness." *Development in Practice* 21, no. 7 (2011): 987–98.

Owen-Smith, Garth. *An Arid Eden*. Illustrated edition. Johannesburg: Ball Jonathan Pub, 2011.

Parks, Thomas. "The Rise and Fall of Donor Funding for Advocacy NGOs: Understanding the Impact." *Development in Practice* 18, no. 2 (April 1, 2008): 213–22.

Prince, Russell. "Policy Transfer as Policy Assemblage: Making Policy for the Creative Industries in New Zealand." *Environment and Planning. A*, Environment and Planning A, 42, no. 1 (2010): 169–86.

Rabinow, Paul. *Anthropos today: reflections on modern equipment*. In-formation series. Princeton, NJ, Princeton, N.J.: Princeton University Press, 2003.

Ramutsindela, M. *Parks and People in Postcolonial Societies: Experiences in Southern Africa*. Springer Science & Business Media, 2004.

Ramutsindela, Maano. *Transfrontier Conservation in Africa: At the Confluence of Capital, Politics and Nature*. CABI, 2007.

Ranta, Eija. "State Governance and Micropractices of Power in the Process of Decolonizing the State in Bolivia." *Forum for Development Studies* 45, no. 3 (2018): 1–23.

NACSO. "Registered Communal Conservancies," 2019.

Reno, William. *Warlord Politics and African States*. Lynne Rienner Publishers, 1998.

van Rensburg, Pierre, and Cecilia Tortajada. "An Assessment of the 2015–2017 Drought in Windhoek." *Frontiers in Climate* 3 (2021).

Richardson, Julie. "Wildlife Utilization and Biodiversity Conservation in Namibia: Conflicting or Complementary Objectives?" *Biodiversity & Conservation* 7, no. 4 (1998): 549–59.

Rija, Alfan A., Rob Critchlow, Chris D. Thomas, and Colin M. Beale. "Global Extent and Drivers of Mammal Population Declines in Protected Areas under Illegal Hunting Pressure." *PloS One* 15, no. 8 (2020): e0227163–e0227163.

Roe, Dilys, Fred Nelson, and Chris Sandbrook. *Community Management of Natural Resources in Africa: Impacts, Experiences and Future Directions*. London, UK: International Institute for Environment and Development, 2009.

Rosenau, James. "Governance in the Twenty-First Century." *Global Governance* 1, no. 1 (January 1, 1995a): 13–43.

Rosenau, James N. "Governance in the Twenty-First Century." *Global Governance* 1, no. 1 (1995b): 13–43.

Rückert, Arne. "Producing Neoliberal Hegemony? A Neo-Gramscian Analysis of the Poverty Reduction Strategy Paper (PRSP) in Nicaragua." *Studies in Political Economy* 79, no. 79 (2007).

Saarinen, Jarkko. *Political Ecologies and Economies of Tourism Development in Kaokoland, North-West Namibia*. *Political Ecology of Tourism*. Routledge, 2016a.

———. *Political Ecologies and Economies of Tourism Development in Kaokoland, North-West Namibia*. Routledge, 2016b.

———. *Sustainable Tourism in Southern Africa: Local Communities and Natural Resources in Transition*. Channel View Publications, 2009.

Sassen, Saskia. *Territory, Authority, Rights: From Medieval to Global Assemblages*. Princeton University Press, 2006.

———. "When National Territory Is Home to the Global: Old Borders to Novel Borderings." *New Political Economy* 10, no. 4 (2005): 523–41.

Sassen, Saskia, and Aihwa Ong. "The Carpenter and the Bricoleur." In *Reassembling International Theory: Assemblage Thinking and International Relations*, edited by Michele Acuto and Simon Curtis, 17–24. London: Palgrave Macmillan UK, 2014.

Savage, Glenn C. "What Is Policy Assemblage?" *Territory, Politics, Governance* 8, no. 3 (2019): 1–17.

Schnegg, Michael. "Institutional Multiplexity: Social Networks and Community-Based Natural Resource Management." *Sustainability Science* 13, no. 4 (July 1, 2018): 1017–30.

———. "Lost in Translation: State Policies and Micro-Politics of Water Governance in Namibia." *Human Ecology* 44, no. 2 (April 1, 2016): 245–55.

Schnegg, Michael, Michael Bollig, and Theresa Linke. "Moral Equality and Success of Common-Pool Water Governance in Namibia." *Ambio* 45, no. 5 (2016): 581–90.

Schnegg, Michael, and Richard Dimba Kiaka. "Subsidized Elephants: Community-Based Resource Governance and Environmental (in)Justice in Namibia." *Geoforum* 93, no. Complete (2018): 105–15.

Schumacher, Ernst Friedrich. *Small Is Beautiful: Economics as If People Mattered*. New York: Harper and Row, 1979.

Severino, Jean-Michel, and Olivier Ray. "The End of ODA: Death and Rebirth of a Global Public Policy - Working Paper 167." Center For Global Development, 2009.

Shackleton, Sheona, and Bruce Campbell. *Devolution in Natural Resource Management: Institutional Arrangements and Power Shifts: A Synthesis of Case Studies from Southern Africa*. CIFOR, 2001.

Shikangalah, Rosemary N., and Benjamin Mapani. "Ephemeral River Systems and Their Ecosystem Provisions to the Local Populations: A Review of the Huab and Ugab Rivers, Namibia" 14 (May 7, 2021): 46–62.

Shilongo, Sem M., Morrie Sam, and Amos Simuela. "Using Incentives as Mitigation Measure for Human Wildlife Conflict Management in Namibia." *International Journal of Scientific and Research Publications (IJSRP)* 8, no. 11 (November 24, 2018).

Silva, Julie A., and Alfons Mosimane. "'How Could I Live Here and Not Be a Member?': Economic Versus Social Drivers of Participation in Namibian Conservation Programs." *Human Ecology* 42, no. 2 (2014): 183–97.

Sindjoun, Luc. "L'Afrique Dans La Science Des Relations Internationales: Notes Introductives et Provisoires Pour Une Sociologie de La Connaissance Internationaliste." *African Sociological Review/Revue Africaine de Sociologie* 3, no. 2 (1999): 142–67.

Sjöstedt, Martin. "Aid Effectiveness and the Paris Declaration: A Mismatch between Ownership and Results-Based Management?" *Public Administration and Development* 33, no. 2 (2013): 143–55.

Snively, Stephen W. "Environmental Laws of Namibia: Constitutions, Conservation, and Cheetahs." *Probate & Property* 26, no. 3 (2012): 44.

Steger, Manfred B. *Globalization: A Very Short Introduction.* Fifth edition., Fully updated new edition. Very Short Introductions 86. Oxford: University Press, 2020.

Stiglitz, Joseph E. *Senior Vice President and Chief Economist*, 1998.

Stoldt, Mirja, Thomas Göttert, Carsten Mann, and Ulrich Zeller. "Transfrontier Conservation Areas and Human-Wildlife Conflict: The Case of the Namibian Component of the Kavango-Zambezi (KAZA) TFCA." *Scientific Reports* 10, no. 1 (May 14, 2020): 7964.

Swatuk, Larry A. "From 'Project' to 'Context': Community Based Natural Resource Management in Botswana." *Global Environmental Politics*, Global Environmental Politics, 5, no. 3 (2005): 95–124.

Taylor, Russell. "Community Based Natural Resource Management in Zimbabwe: The Experience of CAMPFIRE." *Biodiversity and Conservation* 18, no. 10 (2009): 2563–83.

Tickner, Arlene B. "Core, Periphery and (Neo)Imperialist International Relations." *European Journal of International Relations* 19, no. 3 (2013): 627–46.

Tieku, Thomas Kwasi. "Theoretical Approaches to Africa's International Relations." In *Handbook of Africa's International Relations*, 1st ed., 11–20. Routledge, 2014.

Tilly, Charles. *Democracy.* Cambridge University Press, 2007.

de Tocqueville, Alexis. "Tyranny of the Majority." In *Democracy in America*, 255–68. New York: Saunders and Otley, 1835.

Toepler, Stefan, Annette Zimmer, Christian Fröhlich, and Katharina Obuch. "The Changing Space for NGOs: Civil Society in Authoritarian and Hybrid Regimes." *Voluntas (Manchester, England)* 31, no. 4 (2020): 649–62.

Tortajada, Cecilia. "Nongovernmental Organizations and Influence on Global Public Policy." *Asia & the Pacific Policy Studies* 3, no. 2 (2016): 266–74.

"Traditional Authorities Act 25 of 2000 - Namibia Legal Database," 2000.

Transparency International. "Transparency International Report (2006) Africa," 2006.

Tremblay, Pascal. "Wildlife Tourism Consumption: Consumptive or Non-consumptive?" *International Journal of Tourism Research* 3, no. 1 (2001): 81–86.

UNWTO. "International Tourism Growth Continues to Outpace the Global Economy | UNWTO," 2020.

Vira, Bhaskar, and Roger Jeffery. "Analytical Issues in Participatory Natural Resource Management." Global Issues Series. Houndmills, Basingstoke, Hampshire; Palgrave, 2001.

Vrasti, Wanda. "Travelling with Ethnography." Research Methods in Critical Security Studies, April 3, 2013.

Walters, William. "Some Critical Notes on 'Governance.'" *Studies in Political Economy* 73 (April 1, 2004): 27–46.

Watts, Michael J. "Antinomies of Community: Some Thoughts on Geography, Resources and Empire." *Transactions of the Institute of British Geographers* 29, no. 2 (2004): 195–216.

Webber, Ashlea, Jodi Baker, Lisa Gaudry, and Larry A. Swatuk. "Water as Threat and Solution: Improving Health Outcomes in Developing Country Contexts." In *Water, Energy, Food and People Across the Global South: 'The Nexus' in an Era of Climate Change*, edited by Larry A. Swatuk and Corrine Cash, 187–206. International Political Economy Series. Cham: Springer International Publishing, 2018.

World Bank. "Gini Index (World Bank Estimate) | Data." Accessed January 21, 2022.

———. "Overview." Text/HTML. World Bank, 2021a.

———. "Overview." Text/HTML. World Bank, 2021b.

———. "World Bank Climate Change Knowledge Portal," 2020.

World Commission on Environment and Development. *Our Common Future*. Oxford; New York: Oxford University Press, 1987.

WWF. "Species List | Endangered, Vulnerable, and Threatened Animals | WWF." World Wildlife Fund, n.d.

Ziai, Aram. *Development Discourse and Global History: From Colonialism to the Sustainable Development Goals*. Routledge, 2015a.

———. *Development Discourse and Global History: From Colonialism to the Sustainable Development Goals*. Routledge, 2015b.

Index[1]

[1] Note: Page numbers followed by 'n' refer to notes.

A. Heffernan, *The Global Politics of Local Conservation*, Environmental Politics and Theory,
https://doi.org/10.1007/978-3-031-24177-2

Printed in the United States
by Baker & Taylor Publisher Services